A GEOGRAPHY OF THE LIFEWORLD

A GEOGRAPHY OF THE LIFEWORLD

MOVEMENT, REST AND ENCOUNTER

DAVID SEAMON

ST. MARTIN'S PRESS NEW YORK

St. Martin's Press, Inc. 175 Fifth Avenue, New York, NY 10010
First published in the United States in 1979
Printed in Great Britain.

ISBN 0-312-32257-7

Library of Congress Cataloging in Publication Data

Seamon, David.
A geography of the lifeworld.

Includes bibliographical references and index.
1. Environmental psychology. I. Title.
BF353.S38 155.9'1 79-481
ISBN 0-312-32257-7

CONTENTS

List of Tables and Figures

Preface 9

Part One: Seeing Anew

1. A Geography of Everyday Life 15
2. Phenomenology and Environmental Experience Groups 20

Part Two: Movement in the Geographical World

3. Cognitive and Behaviourist Theories of Movement 33
4. Habit and the Notion of Body-Subject 38
5. Merleau-Ponty and Learning for Body-Subject 46
6. Body and Place Choreographies 54
7. Implications for Environmental Theory and Design 60

Part Three: Rest in the Geographical World

8. At-homeness and Territoriality 69
9. Centres, Places for Things and the Notion of Feeling-Subject 73
10. The Home and At-Homeness 78
11. Implications for Environmental Theory, Education and Design 86

Part Four: Encounter with the Geographical World

12. Perception and a Continuum of Awareness 99
13. Fluctuation, Obliviousness and Watching 103
14. Noticing and Heightened Contact 108
15. Basic Contact, Encounter and At-Homeness 115
16. Implications for Environmental Theory and Education 121

Part Five: Searching Out a Whole

17. Movement and Rest 131

18. The Triad of Habituality 138

19. Place Ballet as a Whole 143

20. An Education of Understanding: Evaluating the Environmental Experience Groups 153

21. Behavioural Geography, Phenomenology and Environmental Experience 159

Appendix A: Selected Observations from Clark Environmental Experience Groups (September 1974-May 1975) 163

Appendix B: Commentaries on the Clark Environmental Experience Groups 191

Appendix C: Organising an Environmental Experience Group 201

References 211

Index 223

LIST OF TABLES AND FIGURES

Tables

2.1	Characteristics of Group Members	26
2.2	Themes of the Environmental Experience Groups	28
11.1	Aspects of At-Homeness	87

Figures

12.1	Modes of Encounter: An Awareness Continuum (for waking states)	102
13.1	The Place of Obliviousness on the Awareness Continuum	105
13.2	The Place of Watching on the Awareness Continuum	105
14.1	The Place of Noticing on the Awareness Continuum	111
14.2	The Place of Heightened Contact on the Awareness Continuum	113
15.1	The Place of Basic Contact on the Awareness Continuum	115
16.1	The Awareness Continuum as a Whole	122
17.1	Triad of Environmental Experience	132
17.2	The Dialectic between Movement and Rest	132
18.1	The Tetrad of Environmental Experience	138
18.2	The Triad of Openess and Habituality	139

LIST OF TABLES AND FIGURES

Tables

[illegible]

Figures

[illegible]

Dedicated to
Clark University, its students,
staff, and faculty,
especially in the years 1970-7

What is the obvious? It is that which is taken for granted and never spoken of as such; yet, the obvious everywhere and always guides and supports our culture. The obvious is that with which we already agree – the base from which all action, individual and social, proceeds. Since it is never explicitly discussed nor articulated, the obvious is the most difficult to identify, even though in a disguised manner it lies all around us. To uncover the obvious we must take a step back from the assumptions and attitudes that entwine us – Grange (1977, p.136).

PREFACE

This book is an exercise in looking and seeing. It hopes to help the reader become more sensitive to his or her experiences with places and environments. Stopping to talk on the way to the corner store with a neighbour repairing his pavement, feeling sad that a local bakery has closed, adjusting to the fact that the street on which one lives has just been made one-way, getting lost in a new place, driving long into the night in order to reach home and sleep in one's own bed – situations like these are the groundstones of this book. I ask if such experiences point to wider patterns of meaning in regard to people's relationship with place and environment. Do such experiences, for example, say something about feeling responsible and caring for a place? About the essential nature of spatial behaviour? About the relationship between community and place? About improving places so that they might become more liveable environments, both humanly and ecologically?

The central message of this book is that a satisfying human existence involves links with the locality in which one chooses to live. A sense of personal satisfaction as well as a sense of community are both inescapably grounded in place. Much of social science in the last several decades seems to suppose that people are now easily able to transcend physical space and environment because of advances in technology and science. Indeed, the predominant Western life-style today involves a patchwork of isolated points – home, office, places of entertainment, recreation, etc. – all linked by a mechanical net of transportation and communication devices. At the same time, however, great thinkers as well as people on the street speak with varying degrees of articulateness about a growing sense of homelessness and alienation. They speak of people's increased disrespect for places and the natural environment.

A phenomenological perspective indicates that this deepening malaise may have partial roots in the growing rupture between people and place. The so-called 'conquest' of terrestrial space may have been successful technologically and economically, but not humanly. At least *experientially*, it seems that people become bound to locality. The quality of their life becomes reduced when these bonds are broken in various ways.

Understanding the person-place bond has threefold value. First, it fosters in the reader a growing interest in the essential nature of his or her *own* day-to-day dealings with environments and places in which he or she lives and moves. Second, such understanding provides a tool whereby environmental designers and policy-makers might discover a new perspective and approach for tackling projects and plans for specific places and environments. Third, this understanding might serve as a framework around which concerned people living in a specific place can ask questions in regard to how they themselves might make that place a more satisfying human and ecological environment.

The ideas in this book are in only small part mine. I have been helped by the thoughts of many, including Yi-Fu Tuan, Anne Buttimer, Martin Heidegger, Maurice Merleau-Ponty, J.G. Bennett, and the excellent phenomenological work done by researchers in the Department of Psychology at Duquesne University. I'm especially indebted to Edward Relph and his perceptive phenomenological study *Place and Placelessness* (1976b), which the interested reader should study as a complement to the present book.

I am also grateful for the dedication, interest and insights of participants in the environmental experience groups, which are the crux of this book. These people are: Andra Nieburgs, Melissa Schwartz, Joel Fish, David Jacobson, Nancy Alcabes, Lisa Gardner, Steve Schwartz, Patricia Dandonoli, Peggy Chase, Judy Levin, Nancy Goody, Allan Long, Peter Glick, Bill Parker, Evelyn Prager, Tom Wyatt, Rob Weinstein, Jonathan Robbins, Marion Mostovy, Howie Libov, Emily Sekler, Phyllis Rubin, Peter Barach, Linda Jaffe, Linda D'Angelo Logan, Jere Fore, Kathy Howard and Wendy Hussey Addison.

Other people, past and present, have touched this book in various ways: my parents, Katherine Bloomfield, Nancy and Cliff Buell, Stanley Blount, Saul Cohen, Martyn Bowden, Walter Schatzberg, Gary Overvold, Joe De Rivera, Connie Fischer, Gary Moore, Roger Hart, Mary O'Malley, Jeffrey Albert, Debra Berley, Tony Hodgeson, Edward Edelstein, Henri Bortoft, Peter Rothstein, Vincent Cipolla, Andy Levine, Graham Rowles, Mick Godkin, Curt Rose, Paul Kariya, Kirsten Johnson, Marc Eichen, John Hunter and Nigel Thrift.

Of all these people, I must thank four especially: Anne Buttimer, my graduate adviser and close friend, who let me do what I was interested in and had faith I could reach a completion, Wendy Hussey Addison, who believed that the environmental experience groups would work and helped make sure they did; Valerie de'Andrea, who typed much of the final manuscript and gave support and invaluable criticism; my novelist-friend John Maguire, who patiently studied the original dissertation out of which this book arose and pointed to ways in which it might become more alive and readable. My sincerest thanks to all these people, as well as the many others whom I have not mentioned but who have helped on the way.

David Seamon
Geography Workroom
Clark University
28 September 1978

Part One

SEEING ANEW

What is the hardest thing of all? That which seems easiest: to use your eyes to see what lies in front of them – Goethe (cited in Roszak, 1973, p.310).

1 A GEOGRAPHY OF EVERYDAY LIFE

> I love to sit in the sun. We have the sun so often here, a regular visitor, a friend one can expect to see often and trust. I like to make tea for my husband and me. At midday we take our tea outside and sit on our bench, our backs against the wall of the house. Neither of us wants pillows, I tell my daughters and sons that they are soft – those beach chairs of theirs. Imagine beach chairs here in New Mexico, so far from any ocean! The bench feels strong to us, not uncomfortable. The tea warms us inside, the sun on the outside. I joke with my husband; I say we are part of the house: the adobe gets baked and we do too. For the most part we say nothing, though. It is enough to sit and be part of God's world. We hear the birds talking to each other, and are grateful they come as close to us as they do, all the more reason to keep our tongues still and hold ourselves in one place – Robert Coles (1973, p.6).

Why might a geographer be interested in a frail elderly woman describing her daily life in a small isolated village in the mountains north of Santa Fe? Geography is the study of the earth as the dwellingplace of man. It seeks to understand a person's life in relation to the places, spaces and environments which in sum comprise his or her *geographical world*. The friendly sun, the supportive bench, the warm clay bricks, the singing birds – each is an aspect of the geographical world in which the woman finds herself. Wherever we are, be it small as an apartment or expansive as a desert, strange as a distant country or taken-for-granted as a small adobe home, we are always housed in a geographical world whose specifics we can change but whose surrounds in some form we can in no way avoid.

This book explores the human being's inescapable immersion in the geographical world. The focus is people's day-to-day experiences and behaviours associated with places, spaces and environments in which they live and move. The search is for certain basic patterns which epitomise human behavioural and experiential relationships with the everyday geographical world. Why, for example, do people like the old woman express profound attachment for their home place? What is the nature of everyday movement in space? In what ways do people notice and encounter their geographical world?

The topic of concern is *everyday environmental experience – the sum total of a person's first-hand involvements with the geographical world in which he or she typically lives.* What is the underlying experiential structure of everyday environmental experience? Does it encompass certain basic characteristics which extend beyond particular person, place and time? Clearly, the geographical world is intimately joined with other dimensions of living – the person's socio-economic world, his interpersonal and spiritual worlds, the temporal context which places him in a personal and societal history. As a geographer, I recognise these many linkages but limit discussion to the world of geographical experience and behaviour. What is the nature of human existence as it happens in a geographical world? What in most essential form is man as a geographical being?

The geographer's interest in environmental behaviour and experience is not new. Along with psychologists, sociologists, planners and other researchers, geographers in the last few decades have helped establish an interdisciplinary field which has variously been called environmental psychology, psychogeography, human ethology, environmental sociology, research in environmental perception, or *behavioural geography*, as I call it here.[1] Behavioural geography has explored such themes as spatial behaviour, territoriality, place preferences, attitudes towards nature and the physical environment. Its development reflects a strong need felt in both the social sciences and design professions to understand the inner psychological structures and processes which underlie a person and group's environmental behaviours. Behavioural geography attempts to clarify how human behaviour affects and is affected by the physical environment. This work may provide help in improving existing environments and designing future ones, be they bathrooms, homes, streets, parks, shopping malls – even entire towns and cities.[2] Also, behavioural geography may help the student become more sensitive to the roles that place, space and environment have in his or her own daily life.

This book is different from most work in behavioural geography because it makes use of *phenomenology*, a way of study which works to uncover and describe things and experiences – i.e. *phenomena* – as they are in their own terms. 'Phenomenology', writes Giorgi (1971, p.9), 'is the study of phenomena as experienced by man. The primary emphasis is on the phenomenon itself exactly as it reveals itself to the experiencing subject in all its concreteness and particularity.' Phenomenology explores the things and events of daily experience. Keen explains:

> Its task is less to give us new ideas than to make explicit those ideas, assumptions, and implicit presuppositions upon which we already behave and experience life. Its task is to reveal to us exactly what we already know and that we know it, so that we can be less puzzled about ourselves (1975, p.18).

Phenomenology has recently been heralded as a significantly new perspective in behavioural geography, which conventionally begins with a particular theoretical perspective (e.g. territoriality, spatial cognition) and set of definitions and assumptions (e.g. home ground, territory as a function of aggression, cognitive map, spatial behaviour as a function of cognitive image).[4] Phenomenology, in contrast, strives to categorise and structure its theme of study as little as possible. It seeks to understand and describe the phenomenon as it is in itself before any prejudices or *a priori* theories have identified, labelled or explained it. Phenomenology, says Spiegelberg (1971, p.658), 'bids us to turn toward phenomena which had been blocked from sight by the theoretical pattern in front of them'.

In addition, phenomenology strives for a holistic view of the phenomenon it studies. 'Always a relatively full analysis of any phenomenon must include its relation to neighboring phenomena,' writes Fischer (1971, p.158), succinctly expressing the need for phenomenology to place its topic of study in a wider context of meaning. Most conventional work in behavioural geography brings its attention to one limited aspect of environmental behaviour and experience – getting around in a new city, residents' definitions of neighbourhood, wilderness users' images of wilderness. Phenomenology, in contrast, seeks to understand the interrelatedness among the various portions of environmental experience and behaviour. It works to uncover the parts of everyday environmental experience as at the same time it insists that these parts must reveal a larger whole. In good phenomenology, parts reciprocate parts, and parts reciprocate whole: each gives insight into all the others.[5]

Movement, rest and *encounter* are the three primary themes used to reveal the whole here. Chapters on movement explore the role of body, habit, and routine in our day-to-day environmental dealings, while chapters on rest examine human attachment to place. Chapters on encounter consider the ways in which people observe and notice the world in which they live. I argue that these three themes portray in one possible fashion the essential core of people's behavioural and experiential involvement with their everyday geographical world. In addition, I suggest that these three themes give valuable insight into environmental education and design.

The empirical data for this inquiry are a collection of descriptive reports very much in nature like the elderly woman's account above. These reports were gathered in the context of four groups of people who were interested enough in their personal relationship with the geographical world to meet weekly for several months and probe different aspects of their own everyday environmental experience. These participants were asked to explore, for example, their day-to-day movements in space, the meanings that various places in their lives had for them, the ways in which they made attentive contact with the everyday environment in which they lived. Each of these groups, including myself as leader, is called an *environmental experience group.* Participants in the groups are called *group members.*[6]

This book works to demonstrate that many of the theories and concepts which have found favour in contemporary behavioural research may not be in accurate contact with the phenomena for which they claim to speak. This does not mean that phenomenology seeks to negate or destroy research in behavioural geography but rather asks its practitioners to re-examine the theoretical groundings on which they make their claims. Is spatial behaviour really a function of cognition? Is attachment to place really bound up in territoriality? Do people really prefer the places and environments they say they prefer? By asking questions like these, phenomenology helps the behavioural geographer, environmental psychologist and other such researchers to clarify the starting-points from which their work arises and thereby to establish a more perfect correspondence between behavioural theories and the actual fabric of human environmental experience and behaviour.

Notes

1. For overviews of this interdisciplinary field – by a psychologist, anthropologist and two geographers respectively – see Craik, 1970; Rapoport, 1977; Saarinen, 1976; Porteous, 1977. Also see Moore and Golledge (eds.), 1976; Wapner, Cohen and Kaplan (eds.), 1976; Leff, 1977; Kaplan and Kaplan (eds.), 1978.

2. See, for example, Ittelson *et al.*, 1974; Rapoport, 1977; Porteous, 1977.

3. One of the best introductions to the history and methods of phenomenology is Spiegelberg, 1971, especially vol.II, pp.659-99. Also good are Giorgi, 1970; Ihde, 1973; Keen, 1975. Some of the best examples of empirical phenomenology are found in Giorgi *et al.* (eds.), 1971, 1975. For discussions of the relationship between phenomenology and social science see MacLeod, 1969; Zeitlin, 1973.

4. Statements emphasising the value of phenomenology to behavioural

geography include Relph, 1970; Wisner, 1970; Tuan, 1971a; Buttimer, 1974. 1976. Practical application of phenomenology to geographic themes include Dardel, 1952; Eliade, 1957; Bachelard, 1958; Lowenthal, 1961; Heidegger, 1962, 1971; Buckley, 1971; Fischer, 1971; Tuan, 1974a, 1974b, 1975a, 1975b, 1977; Jager, 1975; Moncrief, 1975; Graber, 1976; Relph, 1976a, 1976b; Seamon, 1976a, 1976b; Rowles, 1978. Note that except for Dardel (1952) and Lowenthal (1961), all work until 1970 was written by non-geographers. Critiques of the relevance of phenomenology to geograpy include Entrikin, 1976, 1977; Ley, 1976; Cosgrove, 1978; Gregory, 1978.

5. Edward Relph's *Place and Placelessness* (1976b) is so far the best phenomenological presentation of a geographical whole – in this case, the experience of place (and its experiential opposite, placelessness). Especially valuable is Relph's inside-outsideness continuum on which can be located different modes of place experience.

6. Observations from these groups, arranged by topic, are included in Appendix A.

2 PHENOMENOLOGY AND THE ENVIRONMENTAL EXPERIENCE GROUPS

> Phenomenology begins in silence. Only he who has experienced genuine perplexity and frustration in the face of the phenomena when trying to find the proper description for them knows what phenomenological seeing really means – Herbert Spiegelberg (1971, p.672).

In normal daily existence people are caught up in a state of affairs that the phenomenologist calls the *natural attitude* – the unquestioned acceptance of the things and experiences of daily living (Giorgi, 1970, pp.146-52; Natanson, 1962). The world of the natural attitude is generally called by the phenomenologist *lifeworld* – the taken-for-granted pattern and context of everyday life through which the person routinely conducts his day-to-day existence without having to make it an object of conscious attention (Buttimer, 1976, pp.277, 281). Immersed in the natural attitude, people do not normally examine the lifeworld; it is concealed as a phenomenon:

> In the natural attitude we are too much absorbed by our mundane pursuits, both practical and theoretical; we are too much absorbed by our goals, purposes, and designs, to pay attention to the *way* the world presents itself to us. The acts of consciousness through which the world and whatever it contains become accessible to us are lived, but they remain undisclosed, unthematized, and in this sense concealed (Giorgi, 1970, p.148, italics in original).

Through a change in attitude – the *phenomenological reduction* as it is usually called – the phenomenologist seeks to make the lifeworld a focus of attention: 'the acts which in the natural attitude are simply lived are now thematized and made topics of reflective analysis' (Giorgi, 1970, p.148). An important tool in this reductive process is *epoche* – the suspension of belief in the experience or experienced thing. The phenomenologist attempts to disengage himself from the lifeworld and re-examine its nature afresh in *epoche* – 'to bring. . .precognitive "givens" into consciousness. . .and enable one to empathize with the worlds of other people' (Buttimer, 1976, p.281).

Epoche does not mean that the phenomenologist rejects the world or his experience of it. Rather, he begins to question these things, as well as all concepts, theories and models designed to describe and explain them. If he conducts *epoche* properly, he may discover that many events and patterns which he previously 'knew' become questionable, while facts that he had previously ignored or deemed insignificant emerge clearly and demand examination and description (Zeitlin, 1973, p.147).

Phenomenology is therefore a descriptive discipline. It attempts to question radically the taken-for-grantedness of lifeworld and theories developed to depict it. Through *epoche*, the phenomenologist looks at human experience anew and records resulting discoveries as accurately as possible.

Group Inquiry as a Phenomenological Method

In traditional epistemologies, modes of knowing have been labelled as either subjective or objective. The focus of subjective knowledge is generally said to be *private* individual experience. In contrast, objective knowledge is usually associated with generalisations and hypotheses which can be tested and replicated *publicly* (Buttimer, 1976, p.282; Rogers, 1969). Phenomenology is a way of knowing that accepts the validity of both traditional modes but is identical to neither (Buttimer, 1976, p.282). Through intersubjective verification – the corroboration of one person's subjective accounts with other persons' – phenomenology attempts to establish generalisations about human experience. Unlike objective modes of study, which seek explanations and causes, phenomenology attempts only descriptive clarification of phenomena and events. It endeavours to 'elicit a dialogue between individual persons and the "subjectivity" of their world' (Buttimer, 1976, p.282).

Group inquiry is one means of fostering this dialogue. Several people interested in better understanding a particular phenomenon meet together regularly for a longer or shorter period of time. They share relevant experiences. They assume that over time this interpersonal sharing and corroboration will lead them as a group to a deeper, more thorough understanding of the phenomenon.

In actual practice, group inquiry has had limited use in phenomenology. One of the few phenomenologists to use it is Fischer (1971), who included it as one technique among several in a phenomenological study of privacy. Though he preceded the formal development of phenomenology by almost a century, the poet and dramatist Goethe (1749-1832) was keenly aware of the value of interpersonal exploration of phenomena and made some use of the technique in his experiential studies of light and colour (Seamon, 1978a, 1978b). Different people are sensitive to different aspects of a thing, Goethe argued, and thus one can discover more about that thing more quickly if his investigatory efforts incorporate the observations of others:

> As soon as one directs the attention of alert, astute individuals to certain phenomena, one finds that they have both predilection for and skill in observation. I have often noticed the fact in my zealous study of the science of light and color, since I frequently discuss the subject of my current interest with persons ordinarily not accustomed to such observations. As soon as their interest is stimulated, they notice phenomena with which I in part was unacquainted and in part had overlooked. In that way they rectify my prematurely formulated ideas, thus giving me the opportunity to advance more rapidly and to emerge from the limitations beleagering one during a laborious investigation (Goethe, 1952, pp.221-2).

The process of group inquiry works to establish a supportive context in which people can build on each other's insights and come to moments of discovery in which unrelated bits of information suddenly fuse together in larger significance, revealing a pattern which was unseen before (Bortoft, 1971; Seamon, 1979). Participants' attention is trained on a thing which is poorly known or only known in *a priori* terms. The aim is to clarify the nature of the thing, attempting to set aside all former conceptions and seeing afresh by the help of each others' observations and insights.

Group inquiry fosters complementary results: descriptive accounts of the phenomenon for the researcher organising the study, and deepened understanding for the participants. The researcher gains experiential data which can be examined and ordered to provide a clearer portrait of the phenomenon. The participants learn to separate from the natural attitude and explore aspects of the lifeworld as objects of attention. They probe aspects of daily living that were taken for granted and less noticed before. Ideally, they become more sensitive to the lifeworld.

The double value of group inquiry and similar phenomenological methods has been described by von Eckartsberg, who depicts the process as a researcher-participant dialogue:

> The dialogue between researcher and researched involves both a change and a learning process although the intentions and foci are different on the part of the researcher, and the researched. On the part of the researcher it is a deeper understanding of the phenomenon under study as lived in action and experience by concrete and hence unique human beings which is of concern, and for the researched the change is in terms of a reflective deepening of understanding of his own living in one of its aspects (1971, p.78).

Possible Problems

Group inquiry involves potential problems. First is the question of generalisation: can broad claims about experience be made on the basis of reports from small numbers of people whose socio-economic and cultural backgrounds are bound to be limited in range? How can such groups speak for a wider human population? Wouldn't an adequate phenomenological study require people from a broad spectrum of lifeworlds?

Conventional scientific methods require statistically proper procedures as a prerequisite for legitimate generalisation. Groups used for study are subject to carefully defined sampling requirements. Phenomenology, in its acceptance of interpersonal corroboration, assumes a different measure of accuracy and objectivity. Phenomenologically, one person's situation speaks for the human situation at large. What is reported about experience in a small group of limited composition may genuinely reflect patterns of experience which have bearing on a wider human sphere.

At one level of human existence, each of us is unique – affecting and affected by cultural, economic and other similar groundings of life. At another plane of existence, however, we each share certain common characteristics. We have four limbs. We move and rest. We more or less have access to the same five sense modalities (cf. Lowenthal, 1961; Tuan, 1974). At this level of study, participants in group inquiry are typical human beings. Their experiential descriptions reflect human experience in its typicality. Certain themes and characteristics pointed to in one participant's experience may find echoes in other people's experience. The aim of the group process is to search out these thematic commonalities and explore them with as much precision as possible.

The question of accuracy is a second problem of group inquiry: are experiential descriptions provided by participants reliable? Do reports genuinely depict the events they purport to describe? Von Eckartsberg

(1971, p.72) has called this difficulty *selective attention* – people experience themselves in their interaction with a particular situation, but they do this selectively, filtering out particular aspects of the experience which they intentionally block from view or don't notice. Because of this difficulty, many social scientists have hesitated to use experiential accounts or have constructed some kind of methodological tool – lie-detector, questionnaire, semantic differential – to convert these qualitative descriptions into some quantitative form acceptable to pre-defined criteria of validity.

In supposing, however, that selective attention is a problem, one also supposes that he or she can arrive at a complete, objective portrait of environmental behaviour and experience. This claim, so often unquestioned in behavioural research, derives largely from a scientific stance that assumes human behaviour and experience to be completely describable and explainable. What, on the other hand, if selective attention is not viewed as a problem, but as a basic characteristic of human nature – that the reach of human awareness is partial and can only elucidate a *portion* of each experience in which we partake? As Von Eckartsberg writes,

> Each event is given in inexhaustible life-process-richness of which we as individuals can become aware only in limited aspects in conscious experience. The reach and focus of our consciousness is limited and elucidates only certain aspects of the totality-event-process in the situation (1971, p.76).

An individual's description of experience may be limited in range and understanding, yet it is an important 'first trace of the experienced event' (ibid., p.76). Group inquiry gathers accounts on experience provided not by just one person but by several people who are willing to consider and question each other's reports. In this sense, the method extends the limited descriptive powers of the single individual to a wider base of intersubjective corroboration and critique. Different people's descriptions highlight different aspects of the phenomenon. Out of the sum arises a composite picture which is greater than each description alone.

The Environmental Experience Groups

Worcester, Massachusetts, is a New England industrial city of some 160,000 persons. Visitors call it dirty and unpleasant. 'Why would you want to live there?' most of them say. Worcester is perhaps best symbolised by the triple-decker, a housing type that looks like a top-heavy shoebox, built in times of economic boom for a flood of new labour. Worcester is a city of immigrants and elderly people. Its streets are twisted and its downtown has a new shopping mall and two short skyscrapers. Worcester is a unique place in many ways, yet at the same time it is a typical human environment in which typical human environmental experience takes place.

Clark University is located in Worcester. It has some 2,200 students and is a small campus – just a few city blocks. Located on Main Street in the heart of a neighbourhood slowly sliding into decay, Clark is surrounded by Worcester and is something of an anomaly. Its image of liberalism – even radicalism – has never been completely accepted by the typical Worcesterite, who speaks of Clark with a curious mixture of affection and distrust.

Worcester and Clark are the setting for the environmental experience groups. All participants except one (an unemployed schoolteacher) were Clark students ranging in age from nineteen to twenty-six (Table 2.1). These people were originally contacted through posters, letters and word of mouth. They received no academic credit for the project. During two semesters, enough people volunteered for the organising of four groups – two beginning groups in the fall and one advanced and one new beginning group in the spring. In total, nineteen persons – ten women and nine men – had some active part in the groups and offered at least a few observations. Some of these people dropped out and others came only irregularly. A core of seven people, including myself, met regularly throughout the two terms on Wednesday and Thursday evenings, usually for an hour and a half. We sat around a small wooden table in a Clark professor's office. The atmosphere was informal and friendly. The aim was to create a situation where people felt free to share experiences and memories with as much clarity and accuracy as possible.[1]

I tape-recorded each meeting, made transcriptions, and then sent copies by campus mail to group members, who could review and reflect on observations.[2] Weekly, too, I sifted through all past and present transcriptions, allowing their contents to reverberate – seeking out thematic linkages which were partial or unnoticed before. By the tenth week, I realised that most reports could be well described in terms of movement, rest and encounter. At the same time, I became aware of important supplementary themes – e.g. the role of body in everyday movement, the emotional link between person and place.

The most difficult problem faced by the groups was the choice of

Table 2.1: Characteristics of Group Members

	Age	Father's Work	Mother's Work	Places lived in and years	Class*
1	23	Department store merchandiser	Housewife	Cleveland, OH (18) Baltimore, MD (2) Ann Arbor, MI (2) Worcester, MA (1)	M
2	20	Elementary school-teacher	College student	Fresno, CA (5) Fremont, CA (14) Worcester, MA (1)	LM
3	21	Sales manager	Office manager	West Orange, NJ (18) Middlebury, VT (2) Worcester, MA (1)	M
4	19	Professor of engineering	Special education teacher	Tenafly, NJ (18) Worcester, MA (1)	UM
5	20	Grocery store manager	Librarian/ secretary	Shrewsbury, MA (5) Needham, MA (13) Worcester, MA (2)	M
6	19	Lawyer	Secretary	Alexandria, VA (13) Arlington, VA (4) Worcester, MA (2)	UM
7	19	Advertising executive	Housewife	Philadelphia, PA (17) Worcester, MA (2)	M
8	21	Lumber dealer	Secretary	Loveland, CO (18) Boulder, CO (2) Worcester, MA (1)	M
9	20	Business executive	Housewife	Spring Valley, NY (18) Worcester, MA (2)	UM
10	19	Chemical engineer	Housewife	Metuchen, NJ (17) Worcester, MA (2)	M
11	23	Unemployed; previously advertising	Law student	Huntington, NY (18) Delaware, OH (2) Worcester, MA (3)	M
12	20	Salesman	Psychologist/ counsellor	Philadelphia, PA (18) Worcester, MA (2)	UM
13	19	Accountant	Teacher	Brockton, MA (18) Worcester, MA (1)	M
14	26	Garage mechanic	Housewife	Richfields Spa, NY (18) Albany, NY (4) England (1) Worcester, MA (3)	LM
15	21	Supermarket manager	Housewife	Bridgeport, CN (18) Worcester, MA (3)	UM?
16	21	Social services	Housewife	Silver Spring, MD (18) Worcester, MA (3)	UM

Table 2.1 *(contd.)*

Age	Father's Work	Mother's Work	Places lived in and years	Class*
17 19	Dentist	Housewife	Teaneck, NJ (18) Worcester, MA (1)	UM
18 21	Attorney	Housewife	Highland Park, IL (19) Worcester, MA (2)	UM

*M = middle class, UM = upper middle class, LM = lower middle class.

focus for weekly discussions. The phenomenologist faces a dilemma in doing phenomenology: he strives not to pre-judge the thing, yet in order to study it, he must organise some guidelines, some pathway by which he can explore the thing. Otherwise, his efforts will lose direction. He will be unable to distinguish essential aspects of the thing from the non-essentials. He must find some middle point which provides guidance for his study yet at the same time allows the phenomenon to be itself.

In the environmental experience groups, I attempted to minimise this dilemma by use of a weekly theme. At the end of each meeting I gave a topic for the following week – e.g. moving in everyday space, emotions in relation to place, destinations, noticing things in the environment (Table 2.2).[3] I asked group members to keep the theme in mind throughout the week and to (1) make observations on any particular experience which might be relevant; (2) reflect on past experiences in terms of the themes. Participants would report their discoveries at the weekly theme meeting. Observations would be given in the order that group members felt like speaking, and it was not mandatory that people speak at all. It often happened that one report would remind someone else of a similar experience, which in turn struck responses in other members. Out of these unpremeditated sequences would frequently arise connections and patterns which were vague or unnoticed before.

I proposed themes in the first few weeks that had arisen out of a previous detailed phenomenology which I had done of my own everyday environmental experience.[4] As the groups proceeded, certain past themes resurfaced and new ones appeared – e.g. places for things, the significance of habit and routine. These thematic patterns were sketchy and fragmented at first. In time, they became clearer and suggested linkages with other themes. Often these patterns became themes in their own right. Of the seventeen themes that we used, I had

Table 2.2: Themes of the Environmental Experience Groups*

1.	Everyday movement in space
2.	Centring
3.	Noticing
4.	Moving in space
5.	Our attention as we move through space
6.	Emotions related to place
7.	A place for everything, everything in its place
8.	Deciding when to go where
9.	Off-centring
10.	Destinations
11.	Care and ownness
12.	Disorientation
13.	Obliviousness and immersion
14.	Paths – attachment to and points along
15.	Order
16.	Spring
17.	The tension between centre and horizon

*A detailed description of each of these themes appears in Appendix C.

decided on five – everyday movement patterns, centring, noticing, moving in space, emotions relating to place – before the groups began. The other twelve arose from the group process itself.

Over time, these seventeen themes produced over 1,400 observations ranging in length from sentence fragment to paragraph or more. As I have said, these many observations eventually separated out under the three main themes of *movement, rest and encounter.* These three themes are the main organisational structure of the book and the reader is advised to establish them clearly in his or her mind.

Adequacy and Use of Text

The adequacy of the group technique used here is dependent on the effectiveness of the following phenomenology. Do its descriptive generalisations apply to most human situations? Can the reader find aspects of his or her own daily life here? Though this phenomenology is based on a limited set of experiences, the argument is that the patterns and linkages discovered apply to other lifeworlds past, present and future. If the groups were conducted in other contexts – with

Sudanese villagers, Pennsylvania Amish, New York sophisticates, or characters in Thomas Hardy's novels – the specific experiential reports would describe a significantly different lifeworld, but underneath should appear the same underlying experiential structures.

Phenomenology is as much a process as a product: the moments of discovery are as significant as the written artifact that describes the discoveries and makes them accessible to others. Sometimes these moments occurred in the group context, with one, a few, or all group members realising the discovery. At other times, insights came to me alone – as I shuffled through observations, sat reflecting, attempted to write, or even walked down the street.

The chapters which follow articulate *my* understanding of discoveries made through the group process. I cannot hope to speak for other group members – each understood different things to different degrees at different times. Appendix B, describing participants' commentaries on the group process, gives some indication of what the group meant for others.

In using the chapters which follow, the reader benefits himself most if he seeks echoes of the presentation in his experience and the experience of other individuals and groups with which he is sufficiently familiar. 'We must. . .', says Grange (1977, p.142), 'see our world exactly as we experience it rather than as we construct it through our rational presuppositions and socialized modes of consciousness.' The aim, in other words, is not to think about the discoveries of the group process – to argue their validity logically – but to search out their existence *in day-to-day experience.* In this way, the reader touches the experiential source of the group discoveries and accepts or rejects them in terms of his own and others' daily living. 'By grasping our existence', says Wild (1963, p.20), 'we can understand thought, but by thought alone we shall never understand existence.'

Notes

1. Appendix C provides instructions for organising an environmental experience group.
2. The tape-recorder made some group members nervous the first few weeks. In time, however, people forgot about its presence and spoke normally and openly.
3. Detailed descriptions of these themes are given in Appendix C.
4. This report is available from the author on request.

Part Two

MOVEMENT IN THE GEOGRAPHICAL WORLD

I shall be passing here this day fortnight at precisely this hour of five-and-twenty minutes past seven. My movements are as truly timed as those of the planets in their courses
– Thomas Hardy (1965, p.24).

3 COGNITIVE AND BEHAVIOURIST THEORIES OF MOVEMENT

> It is darker in the woods, even in common nights than some suppose. I frequently had to look up at the opening between the trees above the path in order to learn my route, and, where there was no cart-path, to feel with my feet the faint track which I had worn. . . Sometimes, after coming home thus late in a dark and muggy night, when my feet felt the path which my eyes could not see, dreaming and absent-minded all the way, until I was aroused by having to raise my hand to lift the latch, I have not been able to recall a single step of my walk, and I have thought that perhaps my body would find its way home if its master should forsake it, as the hand finds its way to the mouth without assistance – Thoreau (1966, pp.113-14).

Movement is an enduring phenomenon in nature. At all scales in the natural world, things and living forms are involved in constant or periodic motion. Continents are slowly displaced by interior earth forces; bulks of soil and rocks are moved by the action of water, wind, and gravity; seeds are transported far from their place of origin; flocks of birds migrate long distances in time with the seasons.

Movement has long been a major theme in geography. Geographers have studied such diverse phenomena as the motion of depositional materials in rivers, the flow of freight over the world's oceans, the spread of domesticated plants and animals from continent to continent. Spurred on by the behavioural perspective, geographers have grown increasingly interested in movement as it occurs at the level of the individual person. Generally, this work has been conducted under the themes of 'activity spaces', 'time geography' or 'spatial cognition and behaviour'.[1]

Movement is taken here to mean *any spatial displacement of the body or bodily parts initiated by the person himself or herself.*[2] Movement was frequently mentioned in the environmental experience groups. Often, discussion focused on movements in the outdoor environment – for example, driving home from work, taking a bus downtown, walking to a shop. Just as frequently, observations described smaller-scaled movements, such as going from one room to another, turning on a light, reaching for a stapler on the desk. Because of these many observations, I eventually made movement a first pivotal theme. Chapters 3-7 explore several characteristics of movement – its habitual nature, its dependence on the body, its types of extension in space and time. When these characteristics are integrated, I argue, they point toward a basic experiential structure of movement – that is, its essential nature as an experience.

This picture of movement will do much to bridge the gap between cognitive and behaviourist theories – the two major ways in which everyday movement in space has conventionally been viewed in social science. Closely associated with the philosophical tradition of rationalism, theories of *spatial cognition* (which most geographers have come to accept) argue that spatial behaviour is dependent on such cognitive processes as thinking, figuring out and deciding.[3] In practice, most of this research has studied a particular individual or group's cognitive representation of space, which is elicited by such devices as map drawings or questionnaires. The assumption is made that a study of these *cognitive maps* (as most geographers have come to call them) will lead to an understanding of the individual and group's behaviour in space. As Downs and Stea explain, 'underlying our definition [of spatial cognition] is a view of behavior which, although variously expressed, can be reduced to the statement that *human spatial behavior is dependent on the individual's cognitive map of the spatial environment'* (1973, p.9, italics in the original).[4]

Alternately, spatial behaviour has been discussed in terms of *behaviourism*, a way of psychology which is linked with the philosophical tradition of empiricism. This perspective views everyday movement in terms of a stimulus-response model – i.e. a particular stimulus in the external environment (e.g. the ringing of a telephone) causes a movement response in the person (the hearer gets up to answer it). In attempting to imitate the methods of natural science, behaviourists have generally restricted their research focus to visible behaviours which can be verified through some form of empirical measurement. They discount all inner experiential processes (e.g. cognition, emotion, bodily intelligence) because they argue that these phenomena are subjective, imprecise, and only knowable by the particular person who reports them.[5] Thus, they study what an animal or person does, rather than what it, he or she experiences. In practice, behaviourist work discussing spatial behaviour as an explicit theme has generally studied rats learning to move through mazes; research with human subjects negotiating space has been much less frequent.[6]

A major weakness of both the cognitive and behaviourist approaches is their insistence on explaining spatial behaviour through an imposed *a priori* theory. The cognitive theorists assume without question that the cognitive map is a key unit of spatial behaviour, while the behaviourists assume the importance of the stimulus-response sequence. Neither group of researchers has felt it necessary to go to the phenomenon of spatial behaviour as it is an experiential process – as it is an experience in the lifeworld. On faith, these researchers have accepted one theoretical approach or the other, in terms of which they then organise their empirical investigations.

I seek to break away from these two opposing theories and return to everyday movement as it is described as an experience in the reports from the environmental experience groups. I look at movement as it is a phenomenon in its own right – before it has been defined, categorised and explained by either of these two dominant perspectives. On the one hand, I bracket the assumption that movement depends on the cognitive map; on the other, that movement is a process of stimulus-response.

In contrast to the view of the cognitive theorists, I argue that cognition plays only a partial role in everyday spatial behaviour; that a sizeable portion of our everyday movements at all varieties of environmental scale is pre-cognitive and involves a prereflective knowledge *of the body.* In contrast to the behaviourist perspective, I argue that this prereflective knowledge is not a chain of discrete, passive responses to external stimuli; rather, that the body holds within itself an active, intentional capacity which intimately 'knows' in its own special fashion the everyday spaces in which the person lives his typical day. Further, I argue that this bodily knowledge is not a structure separate from the cognitive stratum of spatial behaviour but works in frequent reciprocity with it.

Already, some researchers have suggested that much of the theory underlying research in behavioural geography is out of tune with that behaviour as it happens experientially. Buttimer, for example, has spoken of a 'virtual obsession over cognition and the cognitive dimension of environmental behaviour [in] recent years' (1976, p.291). Tuan extends this criticism when he writes 'it can not be assumed that people walk around with pictures in their head, or that people's spatial behavior is guided by picture-like images and mental maps that are like real maps' (1975, p.213). Discoveries here will substantiate Buttimer and Tuan's criticism of research in environmental behaviour, and have considerable relation to the French phenomenologist Merleau-Ponty, to whom I will refer as the chapters proceed.

Notes

1. On activity spaces, see, for example, Chapin and Hightower, 1966; Adams, 1969; Wheeler, 1972; on time geography, see Hägerstrand, 1970 and 1974; Thrift, 1977; on spatial cognition, see Downs, 1970; Downs and Stea (eds.), 1973; Hart and Moore, 1973; Moore, 1973, 1974, 1976; Moore and Golledge (eds.), 1976; Downs and Stea, 1977; Leff, 1977.

2. This definition, by its phrasing, includes such involuntary actions as blinking, breathing, itching, etc. In practice, we discount these movements, since they are largely instinctual and have little direct bearing on people's experience with place, space and environment.

3. Hart and Moore (1973) provide discussion of the philosophical tradition out of which the cognitive perspective arises. The French phenomenologist Merleau-Ponty (1962) provides one of the most forceful critiques of this 'intellectualist' tradition, as he calls it. There is not one approach to environmental cognition, but several; their range is well portrayed in Downs and Stea (eds.), 1973 , and Moore and Golledge (eds.), 1976. The cognitive approach can be broken into two main subgroups, which in theory at least claim to approach environmental behaviour from differing perspectives. Cognitive theorists grounded in the behaviourist tradition accept a stimulus-response model of environmental behaviour but add cognition as a significant intervening process (Tolman, 1973, originally 1948; Osgood, 1953; Stea, 1976). In contrast are the cognitive theorists who take an 'interactive-constructivist' position arising out of the work of Piaget (Hart and Moore, 1973). These students attack the passive role assigned to the organism in behaviourist models of cognition and argue that the person actively mediates his relationship with the environment (e.g. Piaget and Inhelder, 1956; Hart and Moore (eds.), 1973; Moore, 1973, 1974, 1976).

In practice, both groups make use of similar operational techniques, especially mapping and modelling, and it is often difficult to see how their interpretations, when separated from the theoretical language in which they are grounded, are in any way different. Regardless of their differences in emphasis, both approaches assume that environmental behaviour is a function of cognition. For this reason, I feel justified in identifying both groups as cognitive theorists. For a discussion arguing a significant difference between the two approaches, see Moore, 1973, pp.8-13.

4. *The Image* by Kenneth Boulding (1956) was an early philosophical discussion of the cognitive perspective which had impact in social science. Kevin Lynch's *Image of the City* (1961) was one of the first empirical studies using a cognitive approach. This book spawned a sea of replications, of which none, unfortunately, considered the experiential validity of the image structures to which they gave so much weight (e.g. De Jonge, 1962; Gulick, 1963; Appleyard, 1970; Beck and Wood, 1976a and 1976b). Tuan (1975a) provides one insightful critique of this work.

5. 'Behaviorism', writes Taylor (1967, p.516), 'has attempted to explain behavior of men and animals by theories and laws couched in concepts designating only physical things and events. The attempt is, therefore, to eschew concepts involving purpose, desire, intention, feeling, and so on. Such concepts are held to designate, if indeed they designate anything at all, unobservable things and events, whose locus is inside the organism.' For an attack on the assumptions of behaviourism, see Merleau-Ponty, 1962, 1963; Koch, 1964; Giorgi, 1970. For a critique of the experimental method underlying behaviourist work, see Giorgi, 1971a and 1971b.

Like work in cognition, there is not one behaviourist theory but several. See Taylor (1967) for a discussion of their differences as well as behaviourism's philosophical base.

6. Clark Hull's research is one representative of work on rat behaviour in mazes; see Hull, 1952. During the late 1940s and early 1950s, considerable controversy arose between behaviourists, represented by Hull, and cognitive behaviourists, represented by Edward Tolman (Stea and Blaut, 1973). This debate centred on whether learning was rooted in discrete stimulus-response linkages or in cognitive-mapping processes, and generated 'a flood of experiments with albino rats designed to demonstrate that one law (S-R) or the other (cognitive mapping) was a root explanation for all learning' (ibid., p.52).

4 HABIT AND THE NOTION OF BODY-SUBJECT

> When I was living home and going to school, I couldn't drive to the university directly – I had to go around one way or the other. I once remember becoming vividly aware of the fact that I always went there by one route and back the other – I'd practically always do it. And the funny thing was that I didn't have to tell myself to go there one way and back the other. Something in me would do it automatically; I didn't have much choice in the matter. Of course, there would be some days when I would have to go somewhere besides school first and I'd take a different route. Otherwise I went and returned by way of the same streets each time – a group member (1.1.6).[1]

A habit is any acquired behaviour nearly or completely involuntary. Note the habitual quality of movement in the above observation from the environmental experience groups. The group member uses the same route sequence automatically each day. Path movements happen 'by themselves', so to speak, without intervention of conscious attention. 'Something in me would do it automatically,' says the reporter, 'I really didn't have much choice in the matter.' Other phrases from group observations reflect the same self-acting quality; 'You go and you don't even know it' (1.2.1); 'I did it effortlessly and unconsciously' (1.1.13); 'It just happens' (1.1.7); 'I always want to go the same old rote way' (1.1.2).

Habitual movements extend over all environmental scales, from driving and walking to reaching and finger movements. 'You don't remember walking there – you do it so automatically,' said one group member who had little recollection of the walk that got her to class each day (1.2.2). 'It doesn't register with me that I've headed for the "wrong" place,' explained another group member who had recently switched rooms with a flat-mate yet occasionally found himself going to the old room (1.1.9). A third group member, forgetting to place a clean towel under the sink after taking the dirty one to the laundry, found himself reaching for the towel anyway (1.1.11). A fourth group member caught himself dialling his home phone rather than the number he had planned to call (1.1.12). 'My thoughts will be elsewhere,' he explained, 'and my fingers automatically dial the number they know best.'

The group member unable to remember her walks to class gave one particularly descriptive statement of habitual movement (1.2.2). 'You get up and go', she said, 'without thinking you know exactly where you have to go, and you get there but you don't think about getting there while you're on your way.' The phrasing of this statement in almost poetic fashion points to a kind of automatic unfolding of movement with which the person has little or no conscious contact. She finds herself at the appropriate classroom destination without having paid the least bit of attention to the movement as it happened at the time. She has no recollection of the great number of footsteps, turns, stops and starts that in sum compose the walk from home to school.[2]

Cognitive and Behaviourist Interpretations of Habit

Behaviourists and cognitive theorists have dealt with habitual movement in two contrasting ways. The latter students argue that habitual behaviours are not really habitual; that if the person could see the inner processes directing 'habitual' spatial behaviour, he would discover that he is consciously evaluating the situation at hand and making constant use of his cognitive map:

> Admittedly, much spatial behavior is repetitious and habitual – in travelling, you get the feeling that 'you could do the trip blindfolded' or 'do it with your eyes shut'. But even this apparent 'stimulus-response' sequence is not so simple: you must be *ready* for the cue that tells you to 'turn here'. . .or *evaluate* the rush hour traffic that tells you to 'take the other way home tonight'. Even in these situations you are *thinking ahead* (in both a literal and metaphorical sense) and using your cognitive map (Downs and Stea, 1973, p.10, italics in the original).

In direct opposition, strict behaviourists reject any cognitive process intervening between environment and behaviour.[3] They have consistently emphasised the automatic nature of everyday movement, which they define in terms of *reinforcement* – i.e. any event the occurrence of which increases the probability that a stimulus on subsequent occasions will evoke a response (Hilgard *et al*., 1974, pp.188-207). Applied to spatial behaviour, this principle argues that a successful traversal of space over a particular route strengthens the chances that this route will be used the next time the organism traverses that space. Each time the movement is repeated the responses evoking that particular route are *reinforced* and in time the pattern becomes habitual and thus involuntary.

In proceeding phenomenologically, one must place in parentheses these two contrasting interpretations and ask what habitual movement is as an experience *before* it has been defined in cognitive or behaviourist terms. Through this bracketing procedure, one sees the sensitive role that body plays in much of everyday movement and moves toward a perspective similar to Merleau-Ponty's.

The Notion of Body-Subject

> I was driving to the dentist's office and at one stoplight intersection suddenly found myself turning left rather than going straight as I should have done. Just for a moment I was able to observe my actions as they happened – my arms were turning the wheel, heading the car up the street I shouldn't have been going on. They were doing it all by themselves, completely in charge of where I was going. And they did it so fast. The car was half-way through the turn before I came to my senses, realised my mistake, and decided how best I could get back on the street which I was supposed to be on. At the time of the turn, I was worrying about what the dentist might have to do with my teeth. I wasn't paying attention to where I was going. Of course, usually I *do* turn at that stoplight because I have friends who live up that street and I visit them often (1.1.8).

The habitual nature of movement arises from the body, which houses its own special kind of purposive sensibility. Examine the above observation. Something in the group member acts before he can cognitively act, and this 'something' is a directed action *in the hands*: 'my arms were turning the wheel. . .they were doing it all by themselves. . .' A second group member describes the action of turning on a string light-switch: 'my hand reaches for the string, pulls, and the light is on. The hand knows exactly what to do. It happens fast and effortlessly – I don't have to think about it at all' (1.1.10).

The movements occur without or before any conscious intervention. The group member turning on the light doesn't have to bring the action to mind – it happens 'fast and effortlessly'. Similarly, the driver's conscious attention is on the impending appointment. He is not aware of the movement at hand until the error is partially executed: 'the car was half-way through the turn before I came to my senses [and] realised my mistake.' He explains that his arms were 'completely in charge of where I was going', while the second person says that his hand can find the string 'even in the dark'. *By themselves*, the arms move to meet the situation at hand.

The body as the root of habitual movement is pointed to in other observations. One group member described an independent force in her legs which gets her about: 'You let your legs do it and don't pay any attention to where you're going' (1.2.3). A second group member noted that his hands had a taken-for-granted familiarity with his desk-space, reaching automatically for envelopes, scissors and other needed objects (1.1.13). A third group member described his ability to place letters quickly in their proper mailboxes when he worked in a post office (1.9.7), while yet another group member spoke of the fluidity with which her fingers moved over the piano keyboard as she played (1.9.6).

The body as a source of movement extends to the most basic of gestures. Consider stepping. 'The foot would come down and grip, while at the same time the other foot was releasing and moving forward to find a safe spot on which it could rest,' said a group member describing wading in a stream (1.9.1). 'My feet had trouble getting in tune with their spacing – they just didn't feel right,' explained another group member speaking of some stairs he found uncomfortable (1.9.2). They reminded him of a better constructed flight he had walked up in an art gallery: 'my feet felt at home and moved up them easily, whereas these uncomfortable stairs were difficult to manage – they didn't fit my feet.' Individual steps involve a sensibility in legs and feet, which also know extended movements merging to produce specific activities like wading and negotiating stairs.

Underlying and guiding our everyday movements, then, is an intentional bodily force which manifests automatically but sensitively: fingers hit the proper piano keys, arm reaches for string or envelope, hands together put letters in their proper mailbox, feet carefully work their way over a stream-bed, legs carry the person to a destination. Borrowing the term from Merleau-Ponty (1962), I call this bodily intentionality *body-subject.* Body-subject is *the inherent capacity of the body to direct behaviours of the person intelligently, and thus function as a special kind of subject which expresses itself in a pre-conscious way usually described by such words as 'automatic', 'habitual', 'involuntary' and 'mechanical'.*

Cognitive and Behaviourist Approaches to Body

The body as intelligent subject is a notion foreign to both cognitive and behaviourist theories of spatial behaviour. Both perspectives view the body as passive – as an inert thing responding to either orders from cognitive consciousness or stimuli from the external environment. The possibility that the body could be active at a prereflective level has a place in neither perspective.

Consider the cognitive theorists. They view movement as a function of cognition, by which they mean any situation in which the person consciously attends to movement – i.e. makes it an object of conscious awareness through considering, evaluating, planning, remembering, or some similar cognitive process.[4] These researchers have focused little attention on the actual bodily movements which constitute spatial behaviour. They have directed most of their efforts to the cognitive map as a record of the individual's cognitive knowledge of space. They have emphasised the cognitive process which is assumed to co-ordinate relations between environment and behaviour.

In examining his own drive to work, for example, the cognitive theorist Wallace (1961) assumes at the start that his driving experience is best considered as a set of cognitive operations which are grounded in several 'standard driving rules', on the basis of which he makes a particular driving decision. To understand the relationship between these rules, external environment and driving behaviour, Wallace argues that the driver should be viewed as a 'cybernetic machine' which cognitively screens external information, compares it to driving rules, and then sends out an order to the body for an appropriate behaviour (ibid., pp.286-8). There is no possibility that the body could house its own purposeful integrity or manifest independently of the cognitive orders which Wallace claims the cognitive mapping process sends out to the body. The body as intentional subject is lost sight of as the student's attention is directed to the assumed role of cognition.

On the other hand, behaviourists have emphasised the significance of body in their discussions of spatial behaviour, but they have viewed it as a collection of reactions to external stimuli. If the behaviourist were asked, for example, to describe driving behaviour from home to work, he would argue that it involves a succession of reactions to the shifting sights, sounds and pressures impinging on the driver's external sense organs, plus internal stimuli coming from the viscera and skeletal muscles. These various stimuli call out feet and arm movements of the driver which are reinforced each time a particular driving response successfully gets the driver safely to his destination. This series of stimulus-responses is eventually integrated into a smooth step-wise progression which easily and automatically gets the person from home to work each day (Tolman, 1973, p.28, originally 1948).

Group observations indicate that both the cognitive and behaviourist theories are incomplete. The cognitive description ignores the fact that many movements proceed independently of any cognitive evaluation process; that the cognitive stratum of experience comes into play only when body-subject makes a wrong movement, as, for example, when the dialler *becomes aware* that he is dialling the wrong number, or the driver *realises* that he is making a wrong turn. Otherwise, cognition is directed to matters other than the behaviour at hand, such as the impending visit to the dentist.

Yet the fact that cognitive attention can intervene when body-subject errs points to a first weakness of the behaviourist perspective: that behaviour can involve a cognitive component and thus is more than a simple sequence of stimulus-response behaviours. Furthermore, the notion of the body-subject calls into question the whole concept of stimulus-response, since body-subject is an intelligent, holistic process which *directs*, while for the behaviourists, the body is a collection of passive responses that can only *react*.

The above observations give no indication that movement is a response to things in the external environment. Consider the group member turning on the light (1.1.10). The central theme in his report is the directed way in which the hand goes up: 'The hand knows exactly what to do.' The environmental context here seems almost secondary, and in fact the person explains that the arm can find the string as well in the dark as in the daylight. Similarly, the focus in the wrong-turn report is the hands which do the turning, 'all by themselves, completely in charge' (1.1.8). The tone of this observation points to the hands as an intelligent agent in charge of the situation in their own special way. There is no indication that the body is blindly responding to stimuli in the environment as the behaviourist would assume. Rather, the body acts in an intentional way which tackles the behaviour as a whole and proceeds to carry it out in a fluid, integrative fashion.

The Place of Cognition

Though cognition may not have a primary role in everyday movement, it must be realised that it has some role. There are moments in a typical day when movements lose their automatic, unnoticed quality, and the person becomes aware of them.

One function of cognition has already been noted: a habitual action of body-subject is out of tune with the physical environment and cognition intervenes. Consider a group member disoriented in a remodelled snack bar (1.3.1), or a group member confusing directions on a one-way street (1.3.3). Both explain that correcting their mistake involves the intervention of conscious attention. 'I have to stop, figure out where I am, then go,' explained the first person. 'I said to myself, "What's wrong here?", saw the problem, and quickly turned the car in the right direction,' said the driver. The group member walking into the wrong bedroom and the group member dialling the wrong number made similar observations: 'Once I'm at the wrong room, I'll note my mistake and direct myself to where I should be going' (1.1.9); 'then I'll suddenly notice what I've done and become aware of what I'm going to dial' (1.1.12). Mistaken movements activate cognition, which quickly evaluates the incongruity at hand and redirects behaviour.

Second, cognition can intervene before body-subject conducts a particular movement. The need at hand requires a movement different than usual, and cognition gives the order. One group member walked down three flights in a library whose entrances at each floor were virtually the same (1.9.3). She approached the door of the floor she didn't want and noticed an inertial bodily force set to carry her through. 'I could feel my body moving ahead', she said, 'all set to go in.' At the same time her cognitive attention became aware of the door sign indicating floor level. 'It was only my head that told me not to go through,' she explained. 'It looked at the door sign and said, "That's not where I want to go". . .It was only through some kind of consciousness that I could intervene and do what I wanted.'

A third role of cognition occurs in unfamiliar environments, where conscious attention assumes complete control of movements. Descriptions of behaviour in new places emphasise a mental alertness which actively scrutinises environment and gives directions. 'You have to be constantly awake,' said one group member, '. . .looking, searching out the place you want' (1.7.1). 'You have to be "on your toes",' said another group member, 'figuring out if you're on the right street, if you've gone past the house you're looking for, if the house you want is on the right or left' (1.7.2).[5]

Cognition expends more energy than body-subject. 'All that constant watching', said one group member, 'takes a lot of energy. Once you know how to get to a place, it's so much easier. You just go there without having to exert yourself or figuring out where you're going' (1.7.1). Body-subject over time learns the way and the trip becomes easy and comfortable. Pathways become taken for granted and distances which originally seemed long become reasonable and normal. 'Distances', said one group member, 'seem further when you think about them in your mind, but when you get to know them by going, they seem closer' (1.6.2).[6]

Cognition, in sum, has a role in everyday movement, particularly if that movement is new, novel, or occurring in an unfamiliar environment. A larger portion of movement, however, arises from the prereflective sensibility of body-subject. Having placed cognition in relation to body, one can explore body-subject further, asking how it learns movements and how it has been interpreted by Merleau-Ponty.

Notes

1. Numbers in parentheses refer to the location of the observation in Appendix A, which includes all reports used in the present text. The interested reader can turn to this appendix when he wishes (1) an observation in full, or (2) a comparison of one observation with those of a similar theme. These observations are transcriptions of reports made in meetings of the environmental experience groups. Some changes have been made in observations to improve the flow of the text.

2. For additional observations on habitual movements, see Appendix A, section 1.1.

3. Again, it is important to realise that there is a subgroup of behaviourists who accept the basic stimulus-response model but incorporate cognition as an intervening variable. As I explained in Chapter 3, note 3, however, I consider this subgroup to be of the cognitive tradition.

4. On the meaning of cognition and related terms, see Moore and Golledge, 1976.

5. No doubt cognition plays additional roles in everyday environmental experience which have not been considered here. For example, in his critique of research in environmental imagery, Tuan (1975a) describes five functions of cognitive maps which have not been highlighted: they make it possible to give directions to a stranger; they make it possible to rehearse spatial behaviour in the mind so that we can be reasonably sure beforehand that we will be able to get where we wish to go; they serve as a mnemonic device by which we can memorise locations of places, things or people; they are imaginary worlds that depict goals which may tempt people out of their habitual routines; like a real map, they provide means to organise data.

6. See Appendix A, section 1.6 for additional observations on this changing sense of distance. This theme is worthy of further phenomenological exploration.

5 MERLEAU-PONTY AND LEARNING FOR BODY-SUBJECT

> Consciousness is being toward the thing through the intermediary of the body. A movement is learned when the body has understood it, that is, when it has incorporated it into its 'world', and to move one's body is to aim at things through it; it is to allow oneself to respond to their call, which is made upon it independently of any representation. Motility, then, is not, as it were, a handmaid of consciousness, transporting the body to that point in space of which we have formed a representation beforehand. In order that we may be able to move our body towards an object, the object must first exist for it, our body must not belong to the realm of the 'in-itself' – Merleau-Ponty (1962, pp.138-9).

Merleau-Ponty introduced the notion of body-subject in his *Phenomenology of Perception* over three decades ago (1945), and discoveries from the environmental experience groups suggest in a concrete context what he spoke of in more general, philosophical terms. The central problem of philosophy for Merleau-Ponty is the 'origin of the object in the very centre of our experience' (1962, p.71). He concludes that this centre is the body, particularly its function as intelligent subject. A large portion of his work demonstrates how traditional philosophies and their psychological offshoots have ignored the central role of body in human experience and thus misrepresent the nature of man and his place in the world.[1]

Merleau-Ponty's criticism of the cognitive theorists is their treatment of the body as merely a physical entity upon which consciousness may act by an exterior causality: 'my body has its world, or understands its world, without having to make use of my "symbolic" or "objectifying" function' (1962, pp.140-1). Movements of the body are not directed by this conscious force – the '"symbolic" and "objectifying" function' – but by the body's intelligent connections with the world at hand:

> My flat is, for me, not a set of closely associated images. It remains a familiar domain round about me only as long as I have 'in my arms' or 'in my legs' the main distances involved, and as long as from my body intentional threads run out towards it (1962, p.130).

The body has an understanding of the world which is independent of any 'set of closely associated images' which the cognitive theorist would term cognitive map. Movements are learned when the body has understood them, and this understanding can be described as a set of invisible but intelligent 'threads' which run out between body and the world with which the body is familiar. This picture of movement corresponds to group descriptions above – the arms turning the wheel, the legs taking the person to the required destination, the feet carefully choosing a resting place in the stream-bed. The body has within itself the power to initiate these directed movements before and without a need for cognition to screen the world at hand and then implement orders.

In a similar way, Merleau-Ponty questions the behaviourist conception of movement because it also depicts the body as unintelligent – though in a considerably different way from the cognitive approach. The actions of the body in its world 'are not complexes of elementary movements, each "blind" to itself and to the other movements making up the total' (Zaner, 1971, p.153). Merleau-Ponty explains that

> the reactions of an organism are not configurations of elementary movements but gestures endowed with an internal unity. . . Experience in an organism is not the recording and fixation of certain readily accomplished movements. It emerges from aptitudes, that is the general power of responding to situations of a certain type by means of varied reactions which have only their meaning in common. Reactions are not, therefore, a succession of events; they have in themselves an 'intelligibility' (Merleau-Ponty, 1963, cited in Zaner, 1971, p.154).

Merleau-Ponty argues that the body is *active* and that through this activity our needs are efficiently transformed into behaviours. His criticism of the cognitive and behaviourist theories is the same as that presented here: their assumption – arrived at from two opposite lines of reasoning – that the body is 'essentially a passivity in respect of its sensuousness to objects' (Zaner, 1971, pp.157-8).

The body must have within its ken the required habitual behaviours if we are to move our body effectively to meet the requirements of everyday living. Without the structure of body-subject in our human constitution, we would be constantly required to plan out every movement anew – to pay continuous attention to each gesture of the hand, each step of the foot. Because of body-subject, we can manage routine demands automatically and so gain freedom from our everyday spaces and environments. We rise beyond such mundane events as getting places, finding things, performing basic tasks, and direct our creative attentions to wider, more significant life-dimensions:

> There is a freedom from milieu that results from what is stable and pre-established in the subject. This promises to balance out the conventional existential emphasis on spontaneity as the sole reality of freedom, an emphasis that sets freedom and stability in the most radical opposition. . .'It is an inner necessity for the most integrated existence to provide itself with "a hibitual body" if it wishes both to be engaged with the world and to dominate that engagement' (Bannan, 1967, pp.67-8).[2]

Learning for Body-Subject

Body-subject learns through action. Movements become familiar when the body performs them several times and incorporates them into its world of prereflective understanding. One group member who had moved to an unfamiliar part of Worcester reported that she had difficulty finding her way from the new home to work the first few times. After travelling the route for a few days, however, she became intimately familiar with it and could make the trip 'even without thinking about it' (1.8.1). A second group member reported that after a class location was changed, it took him several times before he would automatically go to the new room rather than the old (1.3.2).

Bodily involvement must be active. The movement will not be learned if the body is passive and the movement conducted separately from the body's direct participation. One group member, for example, explained that the past summer she had ridden the same bus route from her home to work each day (1.8.2). When a friend volunteered to drive her one morning, however, she could not give him directions. Body-subject was passively carried to a destination. It had not actively performed the journey and had not learned the pathway.

Body-subject learns through repetition and therefore requires time to familiarise itself with the world in which it finds itself. Once that familiarity is established, body-subject is closely held to it, and by its own initiative is limited in the creation of new routines. This limitation is well illustrated by observations describing confusions of body-subject because of a change in physical environment. The group member paying for his order in the recently remodelled snack bar found himself automatically moving to the former location of the cash register rather than its new place (1.3.1). The group member turning his car left did not notice that the street was one-way and that only a right turn could be made (1.3.3). In both examples, a learned action is out of phase with the world at hand and momentary confusion results.

Body-subject becomes attached to the movements it knows. When the person must conduct a movement different from usual, varying degrees of emotional distress may arise. For example, group members were asked to try an experiment of going to a place by a different route than they usually did.[3] People made reference to a strong feeling of annoyance and dislike when they made or contemplated the change: 'the experiment was an inconvenience' (1.4.2); 'it feels uncomfortable' (1.4.4); 'I found myself consistently not wanting to do this, saying "Why bother?", I didn't feel like going out of my way' (1.4.5); 'I kept putting the experiment off – I didn't want to do it' (1.4.1).[4] Some group members reported a similar feeling of anxiety when they were driven somewhere by a route different from the one they would usually drive themselves: 'I feel a little uncomfortable' (1.5.1); 'I notice myself sometimes getting a little annoyed and anxious, asking myself why this person is going the "wrong" way' (1.5.1); 'I felt uncomfortable because we weren't going the way I thought we should be going' (1.5.3). Body-subject is conservative in nature and prefers that movements adhere to their patterns of the past. This fact has important implications for environmental policy and design.

On the other hand, body-subject has some ability to adapt creatively to new situations. Easily being able to adjust from standard to automatic transmission (1.10.1) or quickly becoming familiar with a larger or smaller car (1.10.2) are examples of body-subject's adaptive powers. For example:

> Driving the larger car felt strange at first. I didn't know how far its sides extended. I noticed that if I didn't worry about it but let the driving happen – just hand it over to my hands on the wheel – they automatically knew what to do, and the driving was easier. Soon it was as if I'd driven the car all my life (1.10.2).

There is a creative power in the hands that properly judges the extension of the larger automobile and safely negotiates through the streets. This adaptability is limited, however, because it is based on repetition.

It must involve the residue of former driving behaviours. Body-subject can not readily adjust, for example, from automatic to standard shift because the required change in habit is too great. The person must practise on the new machine before his movements are an integrated whole again.

The Behaviourist Interpretation of Learning

Repetition is crucial to the behaviourist definition of learning, which is defined as the ability to repeat certain gestures fixed as habits after a period of trial and error (Hilgard *et al.*, 1974, p.189). Behaviourists explain repetition in terms of reinforcement from the external environment (ibid., p.189). Much of their work has attempted to investigate systematically the effect of such reinforcement variables as amount and delay of reinforcement. Do rats receiving a larger food reward learn a maze faster than those receiving less? Do rats receiving a reward immediately after successful completion of the maze learn faster than those whose reward is delayed (ibid., p.189)?

Phenomenologists recognise the importance of repetition in bodily learning, but interpret it as an active endeavour of body by which its powers as subject are extended. This bodily process, as the above observations on driving indicate, can readily adjust to minor changes in its world, but requires time to adapt if the world is considerably changed. Bodily learning is not a sequence of responses established through reinforcement. It is the body's grasping understanding fostered through action.

Merleau-Ponty's criticism of the behaviourist approach is the same as the one here. 'Learning is not the fixation of a particular gesture as a response to a particular situation but rather the establishment of general attitude with regard to the structure or essence of the situation' (Bannan, 1967, p.38). Each situation with which the body is familiar is an analogue to many others, and 'what our experience with them generates are global attitudes, not simply repeatable gestures' (ibid., p.39). Body-subject can transfer its movements over similar contexts because of these global attitudes – from standard to automatic transmission; from smaller to larger cars. Similarly, we can shift our writing strokes from pencil on paper to blackboard, climb a set of stairs we've never used before, or open a door latch which is different from ones we've known in the past.

Generalisation is the term that behaviourists have given to the ability of adjusting to new situations in so far as they are similar to familiar ones (Hilgard *et al.*, 1974, p.194). Again because of their stimulus-response assumption, however, they argue that the cause of generalisation is a stimulus similar to the one which originally evoked a particular behaviour. Drivers are readily able to shift from standard to automatic, behaviourists would contend, because many of the environmental stimuli are still the same – location of accelerator and brake, presence of steering wheel, order of operations. Merleau-Ponty's point is that the power of generalisation lies within the body as subject. Body-subject establishes a general attitude towards particular tasks and to some degree can creatively vary its behaviours.

The Cognitive Interpretation of Learning

Learning, argue cognitive theorists, is a situation in which a particular cognitive structure is elaborated or reorganised to conform better to the world at hand. Unlike the behaviourists, cognitive theorists see the person as active in the learning process: learning, say Hart and Moore (1973, p.250), 'refers to the situation in which information is presented to the individual who changes through reacting to it and corrects initial attempts in response to indications about his prior successes' (also see Hilgard and Bower, 1966; Smith, 1975). To measure environmental learning operationally, these researchers have studied map drawings, verbal descriptions, or toy models over time (e.g. Stea and Blaut, 1973; Klett and Alpaugh, 1976; Beck and Wood, 1976a, 1976b).

Is, however, a person's ability to move in space reflected in his elicited cognitive representation of that space? Map drawings and similar devices elicit cognitive knowledge of the person's geographical world but may say little about his actual ability to move through it. Studies of maps and similar data over time may accurately portray a person's increased cognitive familiarity with his geographical world but may say little about environmental learning – if learning means the actual ability to get around in the world. The causal link between cognition and behaviour is legitimate and directly relevant to an understanding of movement if cognitive mapping is a genuine phenomenon in human experience and behaviour. If, however, the body is the prime source of movement in everyday space, then the relationship between cognition and behaviour is less significant and may not warrant the considerable amount of research premised on it.

Body-subject can manifest knowledge of space only through action. It can not be asked for its geographical knowledge by way of map or interview because its language is comprised of gestures and movements which only 'speak' through behaviours in the moment. What the student can do, as has been begun here, is to examine first-hand accounts of environmental and spatial learning. Consider such situations as a postman's learning a new mail route, a pilot's following a new flight pattern, a boatman's negotiating a strange river. Descriptions of such experiences could be explored for underlying experiential patterns common to many situations in which new movements are learned and eventually become familiar.

Much of the research on spatial cognition is based on the work of Piaget and has studied *development* – 'qualitative changes in the organization of behavior' (Hart and Moore, 1973, p.250). Piaget argues that individuals pass through different developmental stages whereby one's cognitive knowledge of the world becomes increasingly integrated and ordered (Piaget and Inhelder, 1956). Moore (1974) has applied Piaget's stages to cognition of large-scale environments like cities. He depicts a three-stage progression: Level I, a spatial representation that is egocentric and dispersed (present in pre-school ages); Level II, a representation that is partially organised (middle childhood); Level III, a representation that is comprehensive and well ordered (teenage and adult).

This sequence may be valuable in understanding the development of environmental knowing which is cognitive, but one questions its relevance to the learning of actual movements in everyday space, which involve body-subject and have less to do with cognition. Piaget suggests body-subject in his four major periods of development when he speaks of a 'sensorimotor period' which extends from birth to the age of two: 'near the end of this period, the child's behavior may be considered intelligent, although this intelligence is tied to actions and the co-ordination of actions, and does not involve internal representation' (Hart and Moore, 1973, p.260). In focusing its attention on cognition, research on spatial behaviour borrowing from Piaget's four periods forgets the possibility that sensorimotor intelligence may continue to have a major role in spatial behaviour throughout the person's life. The notion of body-subject recaptures its significance, though in different interpretive form.[5]

Notes

1. Good overviews of Merleau-Ponty's work are Kwant, 1963; Barral, 1965; Bannan, 1967. An insightful commentary on the role of body in Merleau-Ponty's philosophy is Zaner, 1971.

2. The significance of body as subject has been recognised by other contemporary scholars – both phenomenologists and not. For reviews of

phenomenological work discussing body see Spickler (ed.), 1970; Zaner, 1971. One non-phenomenologist emphasising the importance of body is Polanyi (1964, 1966). The form and function of the body also have bearing on human experience of the world; see Straus, 1966; Spickler (ed.), 1970; Tuan, 1974b, 1977.

3. See Appendix C, theme four, for details on this experiment.

4. See Appendix A, section 1.4 for additional observations on this experiment.

5. Perhaps one could speak of body-subject in developmental terms, though such a possibility will not be considered here. Why, for example, do some people have a better 'sense of direction' than others? Can other parts of the person, particularly thinking, interfere with the fluidity of body-subject? How developed is body-subject in animals? More or less than in humans? Does body-subject help explain why cats, dogs, and other animals can travel long distances from home and not get lost?

6 BODY AND PLACE CHOREOGRAPHIES

> She has been dusting and sweeping the floor as she talks and now she is finished. Next come the plants, a dozen or so of them; they need to be watered and moved in or out of the sun. . .'I have had them so long – I don't remember the number of years. I know each one's needs, and I try to take care of them the same time each day. Maybe it is unnecessary nonsense, the amount of attention I give. I know that is what Domingo would say. Only once did he put his belief into words, and then I reminded him that he has his habits too. No one can keep him from starting in one corner of his garden and working his way to the other, and with such care. I asked him years ago why not change around every once in a while and begin on the furthest side, and go faster. "I couldn't do it," he said, and I told him I understood. Habits are not crutches; habits are roads we have paved for ourselves. When we are old, and if we have done a good job, the roads last and make the remaining time useful: we get where we want to go, and without the delays we used to have when we were young. . .' – Robert Coles (1973, p.28).

'Habit is the enormous fly-wheel of society, its most precious conservative agent,' wrote psychologist William James (1902, p.121). The root of habit, I have argued, is body-subject, by virtue of which our everyday behaviours can proceed smoothly and automatically. Body-subject is caretaker of life's mundane aspects. A change in its patterns is difficult because, first, the new behaviour must be repeated many times before body-subject learns it, and second, the change may provoke emotional stress.

Body-subject houses complex behaviours extending over considerable portions of time as well as space. These behaviours are of two types which I call *body ballets* and *time-space routines.* When present for many people sharing the same space, these patterns fuse to create what I call *place ballet.*

Body Ballets and Time-Space Routines

Body ballet is *a set of integrated gestures and movements which sustain a particular task or aim.* Body ballets are frequently an integral part of manual skill or artistic talent – for instance, washing dishes, ploughing, house-building, hunting or potting. 'His movements were incredible – they flowed together,' said one group member of a metal-smith at work. 'Both hands were working at once. . .doing exactly what they had to do perfectly' (1.9.4). Operating an ice-cream truck can involve a body ballet. Taking orders, scooping ice cream, making change – all involve a pattern and flow that quickly become routine:

> As I worked I'd get into a rhythm of getting ice cream and giving change. My actions would flow and I'd feel good. I had about twenty kinds of flavors on my truck. Someone would order, and automatically I would reach for the right container, make what the customer wanted, and take his money. Most of the time I didn't have to think about what I was doing. It all became routine (1.9.5).

Basic bodily movements fuse together into body ballet through training and practice. Simple hand, leg and trunk movements become attuned to a particular line of work or action and direct themselves spontaneously to meet the need at hand. Words like 'flow' and 'rhythm' indicate that body ballet is organic and integrated rather than step-wise and fragmentary. Activities require a minimum of cognitive activity: 'I didn't have to think about what I was doing', said the operator of the ice-cream truck. 'It all became routine.'

Similar to body ballet, a *time-space routine* is *a set of habitual bodily behaviours which extend through a considerable portion of time.* Sizeable segments of a person's day may be organised around such routines. Waking at 7.30, making the bed, bathing, dressing, walking out of the house at eight – so one group member described a morning routine that he followed every day but Sunday. From home he walked to a nearby café, picked up a newspaper (which *had* to be the *New York Times*), ordered his usual fare (one scrambled egg and coffee), and stayed there until nine when he walked to his office (1.11.2). 'She is always in a particular place at a particular time and usually doing a particular thing there,' said another group member of her grandmother's daily routine. Between six and nine, for example, the woman is working in the kitchen; between nine and twelve, sewing in the front porch (1.11.1). A third person described her brother's dinner routine on weekdays:

> My brother routinises the things he does at home. For example, he has a dinner routine. He gets home a little after 6.30, puts his briefcase in the dining-room, goes upstairs to change his clothes.

> Then he makes dinner – a salad, a bowl of either canned ravioli or spaghetti, a glass of water. He says he doesn't want to make a choice of menu each day. He eats in front of the seven o'clock news on television (2.9.2).

In time-space routines, a series of behaviours – e.g. bathroom, sewing, cooking routines (in themselves body ballets) – join in a wider pattern directed by body-subject. These routines are not consciously planned but happen naturally. They are taken-for-granted segments of daily living. As the first group member explained, he doesn't figure out his morning schedule each day; rather, 'it unfolds and I follow it.' Like the frustration of going another route, a change in routine can cause irritation: 'I like this routine and I've noticed how I'm bothered a bit when a part of it is upset – if the *Times* is sold out, or if the booths are taken and I have to sit at a counter.'

'Unfolding' describes well the holistic, organic quality of time-space routines. Large portions of a day can proceed with a minimum of planning and decision when a person has established a series of time-space routines in his daily or weekly schedule. The day can 'unfold', so to speak. On the other hand, the person generally becomes attached to these routines, and interference (with seating place, newspaper read, or any other element of the routine) may lead to greater or lesser stress.

Time-space routines automatically appropriate activities through time and are an essential aspect of everyday life. They maintain a continuity in our lives, allowing us to do automatically in the present what we've done in the past. Time-space routines, together with body ballets, manage the habitual, repetitive aspects of life. They free our conscious attention for other more eventful endeavours. On the other hand, time-space routines may be difficult to break or modify because people grow attached to them and forget that life could be otherwise. In this sense, time-space routines are a conservative force. They may be a considerable obstacle in the face of useful change or progress.

Place Ballets

Body ballets and time-space routines mix in a supportive physical environment to create *place ballet* – an interaction of many time-space routines and body ballets rooted in space. The place ballet can occur in all types of environments – indoor, outdoor, streets, neighbourhoods, market places, transportation depots, cafés. The groundstones of place ballet are continual human activity and temporal continuity. Place ballet fosters a strong, even profound, sense of place and has implications for planning and design.

Familiarity arising out of routine is an important aspect of place ballet. One group member, working in a corner grocery, got to know many customers because they came regularly (1.12.1). 'I like it,' she said, 'seeing people I recognise. It helps to pass the time and gives me people to talk to.' The frequenter of the corner luncheonette made the same point:

> A lot of people know each other, and the owner of the place knows every one of the regulars and what they will order. This situation of knowing other people – of knowing who's there at the time, recognising faces that you can say hello to – makes the place warmer. It creates a certain atmosphere that wouldn't be if new faces came in every day (1.12.2).

In place ballet, individual routines meet regularly in time and space. The regularity is unintentional, arising slowly over time as the result of many repeated 'accidental' meetings. People who otherwise might not know each other become acquainted – even friends. At a minimum, there is recognition. Participants generally appreciate the climate of familiarity which grows and to which they become attached. The base of place ballet is body-subject, supporting a time-space continuity grounded in patterns of the past.

Wider Contexts

The notions of body ballet, time-space routine and place ballet are valuable for behavioural geography because they join people with environmental time-space. Though the above examples are limited and culture-bound, their underlying experiential patterns transcend particular social and historical contexts and can be found in all human situations, past and present, Western and non-Western. Consider, for example, the start of a typical day for the Menomini, an Indian tribe living along the northwestern shore of Lake Michigan in the seventeenth century:

> At dawn, the women rose, fetched water, built or rebuilt the fire, and prepared breakfast while the men were getting up. Breakfast was the first of two regular meals per day. The men and boys went off to the hunting and fishing grounds. . .The women worked at home and nearby, tending the crops, processing food, gathering bark and reeds, collecting edible roots and berries, working on clothing,

> weaving mats, and caring for infants and children (Hockett, 1973, pp.13-14).

Time-space routines and body ballets are the foundation of this typical daily pattern; activities follow a sequence which is largely habitual and unpremeditated. The women's activities are an extended time-space routine incorporating many individual body ballets – water-fetching, fire-building, crop-tending and weaving. Each activity requires a particular combination of gestures and movements which correctly manipulate materials at hand and produce the desired artifact or aim. The skill of weaving, for example, is a knowledge of the hands, which long ago learned a proper sequence and rhythm and can now conduct their work quickly and automatically.

One can visualise a series of place ballets unfolding throughout the Menomini's day. The women meet at the stream as they fetch water and partake in conversation. This place is not only a source of water but a scene of community interaction and communication which repeats each morning through the regularity of water-fetching. The underlying structure of the place ballet is no different from the contemporary street scene that Jane Jacobs describes on the block where she once lived in Greenwich Village in New York; note that she calls it a 'ballet':[1]

> The stretch of Hudson Street where I live is each day the scene of an intricate sidewalk ballet. I make my own first entrance into it a little after eight when I put out the garbage can, surely a prosaic occupation, but I enjoy my part, my little clang, as the droves of junior high school students walk by the center of the stage dropping candy wrappers. . . While I sweep up the wrappers I watch the other rituals of the morning: Mr. Halpert unlocking the laundry's handcart from its mooring to a cellar door, Joe Cornacchia's son-in-law stacking out the empty crates from the delicatessen, the barber bringing out his sidewalk folding chair, Mr. Goldstein arranging the coils of wire which proclaim the hardware store is open, the wife of the tenement's superintendent depositing her chunky three-year-old with a toy mandolin on the stoop, the vantage point from which he is learning the English his mother cannot speak. Now the primary children, heading for St. Luke's, dribble through to the west, and the children from P.S.41, heading toward the east (Jacobs, 1961, pp.52-3).

The essential experiential process working on Hudson Street, in the Menomini village, in the corner café, is much the same, though on the surface each place is considerably different. People come together in time and space as each individual is involved in his or her own time-space routines and body ballets. People recognise each other and partake in conversation. Spaces of activity take on a sense of place that each person does his or her small part in creating and sustaining. These places are more than locations and space to be traversed. Each comes to house a dynamism which has arisen naturally without directed intervention.

Place ballet takes on the quality that Relph (1976b, p.55) has called *existential insideness* – a situation in which 'a place is experienced without deliberate and selfconscious reflection yet is full of significances'; people 'know the place and its people and are known and accepted there'. In place ballet, space becomes place through interpersonal, spatio-temporal sharing. Human parts create a larger place-whole. The meaning of the whole is normally expressed indirectly – through day-to-day meetings and an implicit sense of participation. The place ballet becomes an object of participants' explicit attention only when it is threatened by modification or destruction – for example, when a neighbourhood is jeopardised by a proposed expressway, or pedestrian flow along a busy thoroughfare is threatened by street-widening. In these moments, a sense of loyalty may become visible and be sufficiently strong to repulse the threat of change.

Note

1. Although not explicitly phenomenological, the book from which this quotation comes, *The Death and Life of Great American Cities* (1961), is a lucid study of the urban lifeworld. Many of Jacobs's discoveries are applicable to other place contexts and say much about the relationship between environmental experience and physical design. Behavioural geographers have not given Jacobs's work anywhere near the proper amount of study and application it deserves. More will be said about Jacobs as the book proceeds.

7 IMPLICATIONS FOR ENVIRONMENTAL THEORY AND DESIGN

> Subway turnstiles at ten of the busiest stations in New York City have been converted in the last two months to accept quarters instead of tokens, and the switch is causing wide confusion. Hundreds of passengers mistakenly deposit tokens each day and then demand their money back. The token is refunded in a slot at the side of each turnstile, but by the time the bewildered passengers figure this out, rush-hour traffic backs up – *New York Times* (27 April 1978, p.41).

To study the space and environment in which a person typically lives and dwells – his *lived space*, as the phenomenologist sometimes speaks of it – we must recognise that this space is grounded first of all in the body (Bollnow, 1967). Because of body-subject, I know at any moment of my normal experience where I am in relation to familiar objects, places and environments. As Zaner explains, 'It is necessary to conceive of a lived-space, one which is constituted and organized in terms of a corporeal scheme, which is itself constituted by means of bodily movements and actions in specific situations' (1971, p.166).

Explicating the bodily dimension of lived-space provides a picture of stabilising forces within a particular life-style. The student understands how prereflective forces within a particular place continue to make that place what it was in the past. He or she can better imagine the effect of particular environmental or social changes on that stability. If a street is made one-way, for example, will its body ballets, time-space routines and place ballets be helped or hindered? Will a shift in a factory's work hours have impact on the place ballet of the surrounding area?

Consider the change in New York City subway turnstiles. Officials who converted the machines sought to improve transit service. 'We thought we could provide a smoother flow of traffic', said a Transit Authority spokesman, 'and make riding the subway a little more convenient' (*New York Times*, 27 April 1978, p.41). Instead, the change produced longer lines as passengers deposited tokens and then demanded their money back. 'People', said one token seller, 'don't read the sign. Some put a token in and then go under the turnstile; others come back to me. This is the worst week I've had in seven years as a token seller' (ibid., p.41).

Here a simple body ballet is suddenly out of phase with the world at hand. Passengers automatically reach for tokens and place them in the slot; it is not part of their routine to read signs saying, 'Exact change, two quarters only, no tokens.' In time, obviously, people will adjust to the change and service will improve. The point is that officials were not aware of the confusion that might ensue because of the change; had they been, they might have instituted the new machines more carefully, perhaps with advance publicity or announcements by loudspeaker.

Consider a second, more costly, example: an attempt by Los Angeles officials to ease pollution and traffic congestion by turning over one lane in each direction of the Santa Monica Freeway to buses and car pools (Lindsey, 1976). On paper, the proposal seemed reasonable. Citizens, realising the problems of smog and traffic jams, would willingly change their commuting routines and shift to more efficient means of transportation. In practice, the change did not happen; instead, residents assailed the proposal. After five months of legal battle, a federal court ordered the state to return the two lanes to the 'one-man, one-car commuter' (ibid., p.26).

Habitual behaviours can overwhelm a programme that seems valuable, effective and logical. If Los Angeles officials had understood the inertia of habit, they might have better foreseen the consequences of the proposal and modified or rejected it. They would have realised at the start that a threat to commuting routine might generate distress. As it is, money and other resources have been wasted in the unsuccessful venture. Commuters who spent material and psychic energies in voluntarily making the change are now forced to alter their commuting behaviour again. Understanding the bodily stratum of lived-space might have produced a proposal which was more in tune with the commuters' abilities to cope with change.

Relocation of entire neighbourhoods is perhaps the most flagrant and costly case of ignoring routine and bodily stability. Besides being severed from their taken-for-granted social world, relocated residents are removed from a bodily lived-space that was unquestioned before. Some people adjust better than others, but in general one can accept Fried's summation of relocation's impact as he generalises from the experience of displaced residents in Boston's West End:

> Losses [of place] generally bring about fragmentation of routines,

of relationships, and of expectations, and frequently imply an alteration in the world of physically available objects and spatially oriented action. It is a disruption in that sense of continuity which is ordinarily a taken-for-granted framework for functioning in a universe which has temporal, social, and spatial dimensions. From this point of view, the loss of an important place represents a change in a potentially significant component of the experience of continuity (1972, p.232).

Freedom of Choice and the Cognitive Bias

A major weakness of contemporary behavioural geography is its emphasis on cognition to the exclusion of other strata of experience. People are likened to computers: they are seen to evaluate all behaviours cognitively. 'Man is viewed as a decision-maker,' writes Saarinen (1974, p.252), 'his behavior is considered to be some function of his image of the real world, and he is taken to be a complex information-processing system.'

Freedom of choice is the mistaken assumption of the cognitive model. Man is seen to be actively in control of behaviours and able to adapt readily to change. What he decides, prefers, plans for, aspires to is the basis of his behaviour. What he thinks will be what he does. Cognition is an important dimension of environmental behaviour and experience, but it must be balanced by habit and routine. A means of weighing the inertia of body against cognitive factors would help the behavioural researcher to describe better the degree of conflict between stability and change. This approach would provide a better base for predicting the success of plans and policies affecting people's life-styles.

Routine is important to our lives because it releases our more conscious and creative energies from the mundane needs of day-to-day living. Cognition is important because it implements new behaviours and realigns habitual behaviours to shifting needs and changing environments. Cognition, in most general terms, is an innovative force by which people can do other than they have in the past. They can imagine their life being different and then work to make that image a reality.

Freedom arises out of limitation. It must be understood, however, that cognition alone does not generate freedom. Thinking about doing differently or deciding to do differently does not necessarily lead to change. I can make a decision to get up an hour earlier each morning or to drive to work by a different route, but my decision does not necessarily lead to the actual behaviour.

Besides body-subject and cognition, there appears to be some third factor – it might be called *wish* or *need* – that ultimately determines whether a person will be able to overcome life's inertia. Cognition provides us the power to imagine a new pattern, but the relative strength of the third factor decides whether the image will become reality. Consider, for example, the following observation made by ethologist Konrad Lorenz:

> I once suddenly realized that when driving a car in Vienna I regularly used two different routes when approaching and when leaving a certain place in the city, and this was at a time when no one-way streets compelled me to do so. Rebelling against the creature of habit in myself, I tried my customary return route for the outward journey and vice-versa. The astonishing result of this experiment was an undeniable feeling of anxiety so unpleasant that when I came to return I reverted to the habitual route (1966, p.68).

Both body-subject (habitual route) and cognition (Lorenz's realisation of that habitual route) are present in this experience. Lorenz's need to change, however, is weak and he reverts back to the regular route. Pretend, however, that Lorenz decided to change his route because he sought a street less congested or more picturesque. Wish might be stronger in these situations and a different behaviour might happen. Cognition alone does not change actions. We need a thorough phenomenology of the experiential process by which routine is overcome and behaviour becomes different.

The Role of Place Ballet

Place ballet joins people and place in a time-space dynamic and is therefore a useful notion for environmental planning and policy. A strong trend in modern Western society is the fragmentation of places and time into isolated units: home separated from shopping place, neighbourhoods divided by expressways, work segmented from leisure (Toffler, 1970; de Grazia, 1972; Boorstin, 1973). At the same time, social critics speak of growing personal alienation and the breakdown of community (Josephson, 1962; Roszak, 1969, 1973; Slater, 1970; Samuels, 1971). Slater (1970, p.5), for example, defines community as 'the wish to live in trust and fraternal co-operation with one's fellows in a total and visible collective entity'. He argues that community is an essential human desire, but believes that it has been 'deeply and uniquely frustrated' today (ibid., p.5). The result, he argues, is a sense of growing malaise and desperation which in time may lead to severe societal crisis.

Place and *placelessness* are used by Relph (1976b) to articulate fragmentation of community. Places, as he defines them, are 'fusions of human and natural order and are the significant centres of our immediate experiences of the world' (Relph 1976b, p.141). Placelessness, in contrast, is 'the casual eradication of distinctive places and the making of standardized landscapes that results from an insensitivity to the significance of place' (ibid., Preface). Placelessness, Relph argues, arises from *kitsch* – an uncritical acceptance of mass values, or *technique* – the overriding concern with efficiency as an end in itself. The overall impact of these two forces, manifesting through such processes as mass communication, mass culture and central authority, is the 'undermining of the importance of place for both individuals and cultures, and the casual replacement of the diverse and significant places of the world with anonymous spaces and exchangeable environments' (ibid., p.143).

If community and place are essential aspects of a satisfying day-to-day life, then the researcher interested in environmental behaviour can well ask what they are in terms of space and environment. Place ballet has considerable bearing on this question because it brings people together physically and thereby helps foster the 'visible collective entity' of which Slater speaks. Further, it promotes interpersonal familiarity, which is one source of Slater's 'trust and cooperation with one's fellows'. This point is argued by Jane Jacobs, who describes how a 'web of public respect and trust' arises out of unintentional, face-to-face meetings in neighbourhood place ballet:

> The trust of a city street is formed over time from many, many little public sidewalk contacts. It grows out of people stopping by at a bar for beer, getting advice from the grocer and giving advice to the newspaper man, comparing opinions with other customers at the bakery and nodding hello to two boys drinking pop on the stoop (1961, p.56).

Place ballets, argues Jacobs, are the heart of a successful city. The key to their vitality, she believes, is an intimate, close-grained diversity of activities and land use, which mutually support each other both socially and economically and draw people into the streets. The planner Oscar Newman (1973, p.3) takes a similar tack in his notion of *defensible space*: 'a living residential environment which can be employed by inhabitants for the enhancement of their lives, while providing security for their families, neighbors, and friends'. Newman argues that defensible space is more probable in environments where residents recognise each other. He points to the significance of in-passing, unintentional contacts in fostering that familiarity. Newman, like Jacobs, believes that physical design can enhance or hinder a person's sense of place and community. He demonstrates how street patterns, hallway arrangements, physical barriers and other features of environment can be used to facilitate place ballets.

Place ballets are not the only component of place experience. At times, people wish to withdraw, to be alone or with intimates – to separate from the public world of which the place ballet is a part. In addition, place ballets no doubt have impractical and negative aspects, particularly in today's era of social and physical mobility. Does place ballet sometimes generate a sense of provincialism? Is it geared to a time when walking was the main mode of locomotion? Are there some types of people who find place ballets uncomfortable and awkward? Of what value is place ballet for upwardly mobile and professional people whose range of contact is generally not grounded spatially?

These questions notwithstanding, it does seem that place ballet is one essential part of place experience which, if lacking, reduces the meaning of human life. At some time or other, all of us have probably sought out place ballet – to escape loneliness, to watch people, to feel part of a larger human whole, to absorb atmosphere 'of the place' – to extend ourselves beyond our own limited personal world and thereby feel 'more alive'. At the same time, however, the uniqueness of places is faltering, which also means the demise of place ballets. 'The trend', writes Relph (1976b, p.117), 'is towards an environment of few significant places – towards a placeless geography, a flatscape, a meaningless pattern of buildings.'

For the behavioural geographer and other people concerned for place and environment, place ballet is a phenomenon worth saving where it exists, worth fostering in places which are now empty and lifeless but which with change might house place ballet. Later parts of this book will discuss place ballets further, including the question of how environmental design and policy might promote or hinder their happening. First, however, other dimensions of everyday environmental experience must be explored. People do not always move. They and the things of their world, for shorter or longer periods of time, are relatively fixed in space and place. This situation of fixedness is *rest*. It is the theme of Part Three.

Part Three

REST IN THE GEOGRAPHICAL WORLD

All really inhabited space bears
the essence of the notion of home
–Gaston Bachelard (1958, p.5).

8 AT-HOMENESS AND TERRITORIALITY

> Limpets move about and feed when covered with water or while the rocks are still wet. Between tides they tend to cling tightly to the rock, preventing dessication and predation by terrestrial animals. Many limpets seem to adopt a particular spot on the rock to which they return after feeding excursions. This spot may be so worn that the resident limpet just fits snugly into it. This 'homing instinct' has been investigated a number of times ever since Aristotle first reported the habit, but the precise mechanism, the physical basis for the homing ability, and the organs and senses involved are still uncertain – B.H. McConnaughey (1974, pp.263-4).

Rest, like movement, plays an integral part in the processes of nature. Inorganic forms such as rocks and soil remain at rest, relatively fixed in place for most of their lifetimes. Plants are stationary and thrive or succumb largely according to the conditions of their growing place. Rest becomes crucially important for mobile organisms, including man, because it provides a time of inactivity and quiet in which worn parts are repaired and depleted energies restored.

In geography, which is often defined as the study of spatial distributions on the earth's surface, rest is an essential phenomenon because the stationary positioning of things and artifacts is the foundation of all areal arrangements. Geographers in the past have been most concerned with the spatial location and relationships of *tangible* phenomena – for example, the distribution of human populations, the placement of cities, the areal patterning of natural resources.

Recently, behavioural geographers have begun to study rest as it has meaning for individuals and groups. Generally, the guide for this research has been *territoriality* – the theory that animals, persons and groups identify with and defend territories of various spatial extents.[1] This work makes a significant contribution to the understanding of human relationships with place and territory. Discoveries in this book echo research on territoriality, and also explore additional qualities of place which situate the territorial dimensions of rest in a wider experiential structure.

Rest refers to *any situation in which the person or an object with which he or she has contact is relatively fixed in place and space for a longer or shorter period of time.* Participants in the environmental experience groups described experiences of rest in such varied places as city, neighbourhood, house and room. They described regularity of place use and emotional attachment to place.

The essential experiential structure of rest, I argue, is *at-homeness – the usually unnoticed, taken-for-granted situation of being comfortable in and familiar with the everyday world in which one lives and outside of which one is 'visiting', 'in transit', 'not at home', 'out of place' or 'travelling'*. The dwelling-place is generally the spatial centre of at-homeness. At the same time, the person who is at home establishes taken-for-granted places for the things of his everyday life and is familiar and comfortable with a geographical world extending beyond the dwelling-place. The specific physical extent and boundaries of at-homeness are not so much the concern here as the overriding experiential structure which makes them possible. The aim is to identify the essential character of at-homeness, which if not an essential ingredient of people's relationship with place would preclude its manifestation in a particular concrete context, be it rooms, houses, city streets, or whatever.

Place as a Function of Territory

That human and animal relationships with space are grounded in aggression and defence is the key assumption of territoriality.[2] Suttles writes:

> The central theme which is shared by many studies of territoriality is its connection with aggression. Humans along with many other animals kill, maul, and pillage individuals from outside their own territory. At the same time, there are many self-sacrificing loyalties and non-utilitarian exchanges among members of the same territorial group (1972, p.140).

Research in territoriality has defined attachment to place and space largely in terms of fear, protection, exclusiveness and preservation. Consider, for example, Soja's definition of territoriality; note his emphasis on demarcation and distinctiveness:

> Territoriality. . .is *a behavioral phenomenon associated with the organization of space into spheres of influence or clearly demarcated territories which are made distinctive and considered at least partially exclusive by their occupants or definers.* Its most obvious

> geographical manifestation is an identifiable patterning of spatial relationships resulting in the *confinement* of certain activities in particular areas and the *exclusion* of certain categories of individuals from the space of the territorial individual or group (1971, p.19, italics in original).

Territoriality has been applied to human spatial behaviour at a variety of environmental scales, including personal (e.g. Sommer, 1969), urban (e.g. Suttles, 1968), regional (e.g. Soja, 1971) and national (e.g. Gottman, 1973). The main aim of this research has been an understanding of links between the territorial impulse and appropriation of space and place for individuals and groups. As with behaviourists and cognitive theorists, researchers in territoriality rarely question the possibility that their approach could be incomplete or reductive. Territoriality is generally accepted by them as axiomatic to empirical study of geographic behaviour.

Extending Territoriality

I recognise the territorial component of at-homeness but argue that an aggression impulse is only one factor contributing to its nature. As well as a place of protection and defence, the home speaks to the creative, caring part of the person. He who is at home is more likely to live a comfortable existence; he is more likely to extend himself and grow. At-homeness is a prime root of personal and societal strength and growth. It may have a major role in fostering community. For geography and other disciplines of the environment, at-homeness is crucial because it is *the* experience associated with people's resting in a particular place on the earth's surface and proceeding to live there.

The components of at-homeness, as group reports reveal them, are *rootedness, appropriation, regeneration, at-easeness* and *warmth.* In addition, two wider components are explored: first, the habitual, stabilising force of body-subject, second, an emotional stratum of experience – I call it *feeling-subject* – which makes a person's place *his* or *hers.* The forces of body and emotion, in multifold, intertwined fashion, connect the person like invisible threads to the places of the world which he calls home. When the person changes these ties in some way, these forces become stressed and the person experiences annoyance, hostility, confusion, home-sickness, or some similar emotional response.

Notes

1. E.g. Ardrey, 1966; Lorenz, 1966; Hall, 1966; Morris, 1967; Suttles, 1968; Boal, 1969; Metton, 1969; Sommer, 1969; Pastalan and Carson (eds.), 1970; Boal, 1971; Buttimer, 1972; Suttles, 1972; Gottman, 1973; Newman, 1973; Ley and Cybriwsky, 1974; Vine, 1975; Scheflin, 1976. Two excellent overviews of territoriality are Soja, 1971, and Malmberg, 1979.

2. A large portion of the territorial literature has focused on one highly debated question: is territoriality a biologically or culturally conditioned phenomenon in man? Several well known authors such as Lorenz (1966), Ardrey (1966) and Morris (1967) have argued that territoriality is a natural instinct in both animals *and* people. Other students such as Allard (1972) have argued against a biologically determined view, pointing out that though territorial instinct may be present in man, it manifests (or fails to) according to the culture in which the individual finds himself. 'What kind of behavioral system emerges', writes Allard (ibid., p.13) 'must conform to man's biological capacities, but since these are wide, the capacities alone tell us little about the real systems undergoing the selective process.' For other criticisms of the biological view, see Callan, 1970, and Soja, 1971.

9 CENTRES, PLACES FOR THINGS AND THE NOTION OF FEELING-SUBJECT

> One day in my rambles, I discovered a small grove composed of twenty or thirty trees, at a convenient distance apart. . . This grove was on a hill differing in shape from other hills in its neighborhood; and after a time, I made a point of finding and using it as a resting-place every day at noon. I did not ask myself why I made a choice of that one spot, going out of my way to sit there, instead of sitting down under any one of the millions of trees and bushes on any other hillside. I thought nothing about it, but acted unconsciously. Only afterwards it seemed to me that, after having rested there once, each time I wished to rest again, the wish came associated with the image of that particular clump of trees. . .; and in a short time I formed a habit of returning animal-like, to repose at that same spot – W.H. Hudson (cited in James, 1958, pp.167-8).

Wherever we go, even for the shortest periods of time, we establish places around which we orient our world and our spatial activities. These *centres* give the person spatial and place identity; they locate him in the environment where he finds himself. Some group members spoke of visiting a new city, for example, and observing that their lodging place immediately became a centre. 'I quickly got my bearings in terms of the friend's house where I was staying and came and went in terms of it,' said one group member (2.1.2). Another group member, even on his first day of visiting, found himself returning automatically to the place where he was staying. 'I took a bus back to the apartment where I was staying,' he explained, 'I did it without thinking. I could have done anything, but I went back there' (2.1.1).

Centres are established during shorter trips away from home. A car, for example, may become a temporary centre on a shopping trip. 'Especially in a place with which I'm not too familiar', said one group member, 'I've noticed how the car becomes my focus in space and I direct my shopping movements in terms of it' (2.1.6). Even in places dealt with only briefly, such as a roadside stand or transportation terminal, we tend to establish centres and orient ourselves around them:

> When you stop by the side of the road to eat lunch and stay for a

> time resting – even there you pick a place, sit down, and then usually spend the rest of your time in terms of that place. Or when you wait in a bus station for a few hours, you get up for some candy, or go to the bathroom, or take a walk, but then more than likely you'll return to the same seat (2.1.3).

Specific implements and fixtures such as seats, desks, tables and beds become centres in interior space: 'my family *always* sits in the same seats at the dinner table' (2.13.1); 'my desk and the big rocking chair I got from the Salvation Army are the two places where I usually am when I'm in my room' (2.13.4); 'at school, my favourite place is my desk – it seems to be the centre of what I do' (2.14.8). Beyond interior space, larger places such as offices, parks, shops, eating establishments and other foci of activities become centres when the person uses them frequently – for example, going to a park for a daily walk (2.14.5) or visiting a local bakery to buy bread (2.14.4). These places may involve regularity of use and thus be connected with time-space routines and body and place ballets (2.14.1-2.14.8).

Places for Things

Besides establishing centres for themselves, people also establish *places for things.* Each object becomes associated with a particular place, and in this way the person can order living-space in terms of things. At the scale of clothing, for example, particular pockets and pouches provide places for specific items. 'I always keep change, keys, and pens in my right pocket and tissues in the left,' explained one group member (2.16.2). 'I have specific places in my pocket-book for certain things,' said another group member. 'In the pouch in front I have pencils and pens. Inside is a zipper case where I put my keys' (2.16.3).

Likewise, shelves, drawers, cupboards and closets provide places for things and create what Heidegger (1962, p.136) has called *regions* – taken-for-granted totalities of places through which the person quickly locates the various things and utensils required for a particular task or occupation. Regions, says Heidegger (ibid., pp.137-8), have the quality of 'inconspicuous familiarity' because they are generally unnoticed and only come to attention when one fails to find something in its place. Inconspicuous familiarity was well described by one group member who realised the taken-for-granted order of kitchen space:

> Because of the group, I've come to be more consciously aware of how important places are to me in the kitchen. All the things I use

> have definite places – even the spices in my spice rack. When I'm preparing a meal, I can quickly locate ingredients and utensils without having to think about it at all. Everything is at hand and ready for use (2.17.2).

In outdoor environments, people less frequently create places for things, since most exterior elements are fixed in position and not easily moved. Even here, however, some individual at some point in time must originally locate a particular thing – be it planted tree, water fountain, garage or whatever. Often this locating process requires considerable time and may become a significant event for person or family. One group member described a family argument arising over the location of a toolshed (2.22.1), while another described a similar dispute over the planting site of a tree (2.22.2)..

For large mobile objects like cars, place may become very important. Several group members reported that they parked their cars in a regular place, one group member described the annoyance and 'loss' resulting from not having a parking space:

> I've forgotten several times this past semester where I parked my car. I find myself stopping at the first stop that looks convenient and parking there, then when I go out to find the car at the end of the day I can't remember where I've parked. As I look for it, I find myself thinking, 'Where did I park this morning, where would have been the most sensible place to park?' But often this logical approach doesn't work and I just have to go around and look. It's ridiculous and annoying at the same time. For the sake of convenience, I'm beginning to establish a parking place (2.21.2).

The Notion of Feeling-Subject

Annoyance due to misplacing a car indicates an emotional stratum of experience directed to place and space. This affective relationship was intimated in Chapter 5 when reporters spoke of an emotional resistance to change in movements or routines. This emotional stratum is now made a phenomenon in its own right and its significance explored in relation to at-homeness.

A sense of attachment is one manifestation of the emotional linkage between person and place. People speak with fondness of places which are or were important to them (2.14). Similarly, places for things foster attachment, especially if the place has existed for a long time (2.20, 2.21). Attachment is sometimes described in terms of *attraction*: the place seems to draw the user to it. 'A magnetic force draws me there,' said a group member who used a crafts workshop regularly (2.14.3). 'If you go out for a while, you're drawn back,' explained another group member, commenting on the frequency with which she and her room-mates used the kitchen in their flat (2.13.9). Attachment to place is also described by *closeness*: the user feels near the places he likes. 'I feel close to that park,' said a group member who went there frequently to sit and be alone (2.14.5). 'It seems near to me,' said a group member of the bakery where he bought bread regularly (2.14.4).

Attachment to place relates not only to positive emotions; it is also associated with a constellation of negative emotions, including anxiety and annoyance. Negativity most commonly arises when places are changed in some way. Consider a dining-room table put back slightly out of place after cleaning; family members began to feel uncomfortable during dinner, got up, and moved the table to its proper place (2.18.2). 'There's this drawback feeling that wants me to keep it as it is,' said a group member describing a change in furniture she had decided not to make (2.20.1). 'It didn't feel right and I turned it around,' said another group member using a typewriter which was facing in a direction different from usual (2.18.3). Yet another group member expressed mild annoyance because a university snack bar had been moved a considerable walking distance from its former place:

> Since they've moved the snack bar to its new location I feel a little uncomfortable when I go there. It seems somehow wrong walking to the new place when in the past I've gone to the old location. I still go but it seems strange. It will take me some time to get used to it (2.15.2).

The experiential stratum associated with attachment to place I call *feeling-subject.* Feeling-subject is *a matrix of emotional intentionalities within the person which extend outwards in varying intensities to the centres, places and spaces of a person's everyday geographical world.*[1] Feeling-subject works in two ways: it sustains positive feelings for well used centres and places, and expresses negativity when these centres and places are changed in some way.

Feeling-subject houses an intelligent directedness similar to body-subject, but different in the sense that it arises from the emotional rather than bodily part of the person. Feeling-subject, coupled with body-subject, is a primary experiential force underlying our daily relations with the geographical world. Though it speaks in a language foreign to cognition and logical thinking – i.e. affective expression – feeling-subject can be said, like its bodily counterpart, to act intelligently and consistently. Feelings for centres or places may often seem logically incongruous or foolish. Yet as the cliché expresses it, 'The heart has a mind of its own' and acts in a way internally consistent with emotional bonds to place and space (2.3.2, 2.3.3).

Exploring the exact nature of feeling-subject is not attempted here; present group observations are not detailed or precise enough. Clearly, feeling-subject establishes links with other portions of the world – its social, economic, interpersonal aspects. Love of person, art or God; sense of duty; dislike of prejudice – all these experiences have their grounding in feeling-subject. What is needed is a phenomenology of the emotional stratum of human experience.[2]

Suffice it to say, both body- and feeling-subjects require time to become familiar with and attached to new environments. These two forces prefer a minimum of change in their lifeworld and react in confusion and annoyance when the taken-for-granted is threatened or upset. Most people live in lifeworlds which are relatively stable and non-changing. This continuity is closely linked with *at-homeness*, for which the notions of body- and feeling-subjects provide important grips to explore its nature.

Notes

1. Feeling here is synonymous with emotion.
2. Perhaps future phenomenologies of lived-space can provide some insights. Does attachment to place say anything about how feeling-subject develops? Is it accurate in the end to arbitrarily separate body- and feeling-subjects as I have done here? Does emotional attachment ever become a destructive force in the person? What overcomes feeling-subject and allows a person to forsake centres or change places and routines? Why do different people feel different amounts of emotional reaction to the same situation?

10 THE HOME AND AT-HOMENESS

> The moment he entered the Pullman he was transported instantly from the vast allness of general humanity in the station into the familiar geography of his own home town. One might have been away years and never have seen an old familiar face; one might have wandered to the far ends of the earth;. . .one might have lived and worked in the canyons of Manhattan, until the very memory of home was lost and far as in a dream, yet the moment that he entered K19 it all came back again, his feet touched earth and he was home – Thomas Wolfe (1973, p.46).

Home is the most important centre. 'My apartment is my special place' (2.4.8). 'I feel bad about leaving my apartment; I've gotten attached to it and I'm going to miss it' (2.3.1). Attachment for home is sometimes best revealed in out-of-the-ordinary situations such as heating failures (2.3.2, 2.3.3). People may bear cold and threat of sickness because of their bonds with home:

> I remember when the heating system in my apartment was broken for a few days last winter. Friends invited me to stay at their houses but I didn't go. Like New Year's Eve. The friends I was with told me to stay but I couldn't think of doing it. It didn't seem right staying in their place when my apartment was just a few blocks away. It was cold in the apartment, but I wanted to be home. I remember thinking to myself how irrational I was being – that I wouldn't be comfortable and might get sick. The thoughts had no effect. I found myself returning home with no hesitation whatsoever (2.3.3).

Attachment to home is associated with the experience of *at-homeness* – the taken-for-granted situation of being comfortable and familiar with the world in which one lives his or her day-to-day life. Observations on home point to five underlying themes which mark out the experiential character of at-homeness – *rootedness, appropriation, regeneration, at-easeness* and *warmth.* Each theme is discussed in turn, then integrated to arrive at a picture of at-homeness as a whole.

Rootedness

Rootedness is the power of home to organise the habitual, bodily stratum of the person's lived-space. Literally, the home roots the person spatially, providing a physical centre for departure and return. Although inescapably a part of a larger geographical whole, the home is a special place because around it the person organises his comings and goings: 'Space isn't all equal for me. Where I live is a unique place because I'm always leaving it and coming back. In one sense, I'm bound to that place' (2.2.2).

The body is the foundation of rootedness. Through the recurring cycle of departure and return, body-subject comes to know the placement of home and its relative location in terms of paths, places, people and things. Body-subject left to its own devices illustrates this fact well. One group member, driving back from a bus station and busy in conversation with the friend riding with him, planned to drive the friend home. Instead, the driver suddenly found himself returning to his own residence. 'How dumb!' he said, 'Here I am driving us back to my house when I have to take you home' (2.8.2). Another group member, caught up in worry, suddenly found himself walking up the stairs of his apartment rather than going to the nearby post office as he planned (2.8.1). Cognition dwells on matters other than movement at hand, and body-subject directs itself towards home – its most frequent destination.

Bodily familiarity extends within the home, establishing places for things and temporal regularity for activities. The person who is at home can move fluidly through the dwelling because body-subject knows that space intimately. Necessary objects and devices are literally 'at hand':

> My mother knows the exact location of everything in our house; she has a place for everything. She doesn't have to figure out where a particular thing is – she goes to it automatically. Like I'll need some string, and she'll know the right drawer where it is. I'd have to check a few places before I'd find it – if I ever did (2.17.1).

Body ballets and time-space routines are intimately associated with home. Rituals such as waking, grooming, dressing and cooking have a particular routinised time and place within the home (2.9). In the same way, departures and returns may be fixed by habit. Consider, for example, the following morning time-space routine, which ends with departure from home at '8.50 sharp':

> On working days, my father follows the same routine each morning. He automatically gets up at 7 o'clock – he doesn't need an alarm. He puts on some old clothes, goes to the bathroom, then picks up the morning newspaper from the front stoop. He puts two sausages in a pan over low flame. They'll be ready to eat at 8.15. While they cook, he reads the paper, always sitting in the same chair. He slouches. Just before the sausages are done, he softboils an egg; he doesn't even wash the pan but uses the same water day after day. He puts a piece of rye bread in the toaster and pours a glass of orange juice. . . After breakfast – he calls it his 'three-minute breakfast' because that's how long it takes him to eat it – he puts the dishes in the dishwasher, shaves, bathes, dresses, and leaves the house at 8.50 sharp (2.9.1).

A substantial portion of a person's everyday behaviours happen automatically because of rootedness; he thus conserves mental energy. Rootedness is established through physical action and requires time to develop. The person who lives in the same place his entire life establishes rootedness in the first few months and years of childhood; the person who changes places must re-establish rootedness each time he moves.

Partially because of rootedness, at-homeness can not be established at once. Spatial familiarity and comfort can arise only through an active integrative process over time. Eventually, space is no longer a set of objective areas, things and points in terms of which behaviours must be figured out cognitively.It becomes a field of prereflective action grounded in the body.

Appropriation

The home *appropriates* space. Appropriation involves, first, a sense of possession and control: the person who is at home holds a space over which he is in charge. Appropriation is disturbed when a home is infringed upon in some way; feeling-subject immediately reacts. 'I've gotten angry about it,' said a group member in regard to an acquaintance staying in her apartment who had taken the liberty of walking in without knocking. 'He doesn't live there and he violates our sense of privacy when he walks in like that' (2.6.1). Workmen fixing a group member's apartment generated a similar feeling of anxiety:

> For the past few weeks workmen have been renovating the apartment house where I live. I've been trying to observe my reactions to them. There's a feeling of trespass: 'What are these people doing here in my building?' I find their presence annoying. I realise they have to be there, but there's tension in having them there (2.6.2).

Appropriation is partially a function of the home's physical context but, more significantly, involves the resident's ability to control passage in and out. The group members feel uncomfortable because of the *uninvited* entrance of the acquaintance, the *uncontrollable* presence of the workmen. Expected or taken-for-granted entrants are familiar or scheduled and do not upset the sense of at-homeness. Uninvited entrants, however, interfere with the resident's sense of control, and cause feeling-subject to react immediately.

A second aspect of appropriation is privacy. 'He violates our privacy,' says the group member above. A place to be alone is part of at-homeness, and the person whose home does not provide such a place feels a certain degree of upset. 'It felt like the room wasn't mine because he was there so much,' said one group member whose dormitory room-mate rarely left the room (2.5.2). 'With the additional person, it seems like there's always someone there, and that I never have the place to myself,' said another group member who now lived with two people rather than one. 'It's a relief every so often to know that both room-mates will be away and I can have the whole apartment to myself' (2.5.2).

Appropriation, as a whole, relates to the resident's ability to control home-space. Lack of appropriation involves infringement or loss of privacy. In either case, the resident has lost his taken-for-granted powers to control and use his home-space as he sees fit. Disruption of appropriation leads to responses of feeling-subject which may include anger, anxiety or discomfort. This emotional response may be long or short, but while it lasts the person is not fully at home.

Regeneration

The person who seeks rest seeks repose and refreshment. *Regeneration* refers to the restorative powers of the home. Most obviously, the home houses physical rest. Several observations on home made reference to sleep and sleeping place. The group member, for example, who came back to the apartment without heat, returned to sleep (2.3.3). Another group member went without dinner and drove late into the night to be able to sleep in his own home rather than elsewhere. 'There was this irresistible urge,' he said, 'to be in my own place and sleep in my own bed' (2.3.4).

The bed itself is often a special place in the home. Some group members spoke of its inviolate character; they felt uncomfortable sitting on a bed without the owner's explicit permission (2.13.5, 2.13.6). The special significance of bed is partially related to sexual activity: it is a place of mutual love and procreation. In this sense, it is the spatial origin of humankind. In addition, the importance of bed is related to appropriation: without the security and privacy of the home, effective rest would be difficult or impossible. The sleeping person can fully prostrate himself without fear or threat.

Locking rituals point to the importance of security during times of rest and sleep. Some group members explained that locking doors was an integral element of their bedroom routine. One group member described his parents' locking ritual which had followed the same basic sequence for years:

> My mother and father lock the door every night before retiring. Usually my father does it, but then my mother rechecks to make sure he hasn't forgotten. My father has a regular routine. He goes to the outer porch door, flips on the yard light, checks the outdoor thermometer, shuts off the light, locks the porch door, then comes in and locks the inner door which comes into the kitchen. Then about fifteen minutes later, my mother gets ready for bed, and she checks the door too. In the morning, my mother gets up first. As soon as she's downstairs, she unlocks both doors and looks out to see what the weather is like. They've done this for as long as I can remember (2.6.3).

Besides sheltering sleep, the home may also foster *psychological* regeneration. 'I go there to get oriented,' explained one group member (2.4.1). 'I go back there to get myself together before another class,' said another (2.4.3). One group member found herself returning home after a professor expressed dissatisfaction with a paper she had written; it is the simple return home, the report suggests, rather than any specific activities there which renews her energies: 'I found myself walking back to my apartment, just to recuperate. I didn't know what I'd do there, but I knew the apartment would help me to feel better' (2.4.5).

In its regenerative powers, the home provides a stable place in which the person can recoup his physical and psychic energies. The person at home has a place where the possibility of rest is taken for granted and secure. Without a place for regeneration, a person's life almost surely disintegrates.

At-Easeness

At-easeness refers to the freedom to be: the person who is at home can be what he most comfortably is and do what he most wishes to do. 'My home is where I can best be myself,' said one group member (2.4.6). 'It's the place where I can "let my hair down",' said another (2.4.2). The home is a place where impulse can be spontaneous and free. It contrasts with public environments where people must partake in roles and behaviours required to maintain a particular public image. Within limits, the person at home can manifest all sides of himself and fear no repercussions; he can be as foolish, negative, or loving as he wishes:

> My apartment is my special place where I feel that I can do the things I like and not feel bothered or guilty. Reading quietly, sitting with a friend, playing the recorder – all these things I can do anywhere, but somehow they seem best done at home. At home I don't feel ashamed to be miserable. I can go to my room, shut the door, be as ugly as I want. I can be angry with my room-mate and it will be okay. No strings are attached to anything I do at home (2.4.8).

The importance of at-easeness may become especially clear in times of sickness. Sick people, explained one group member, are most comfortable at home because they do not have the energy to 'pretend you're something you're not' (2.4.4). The invalid feels comfortable to be as sick as he is at home; he can be completely vulnerable and not fear the consequences. In contrast, it is this vulnerability which makes the guest uncomfortable in the host's home (2.12). Because the home which he visits is not his own, the guest may feel uncomfortable and ill at ease. Accordingly, etiquette requires the host to reassure and guide the guest, who is often grateful (2.12.1).

At-easeness may be reflected in the home's physical character, which sustains at-homeness as at the same time it helps create it. 'It's good to live in a place that shows by what it is who you are,' explained one group member who had redecorated her apartment and put up wall hangings that 'all tell something about who I am' (2.3.6). 'It's very hard to make a personal mark,' said another group member who had difficulty individualising his dormitory room because of a uniform concrete construction that resisted modifications. 'I don't feel as comfortable there as I might if I could make the place my own' (2.3.5).

At-easeness supports a renewal of self. It is a groundstone from which can develop personal and interpersonal growth. 'Ill at ease' connotes sickness. To be ill at ease in one's own home indicates unnaturalness and leads to resolution or physical and psychological stress.

Warmth

Warmth refers to an atmosphere of friendliness, concern and support that a successful home generates. 'The house had a warm feeling, I felt very good being in it,' said a group member vividly struck by his first visit to a friend's house (2.7.4). 'The place felt so much a home,' said another group member. 'It felt so warm and cosy. I almost wished I was a child living there – it felt so supportive' (2.7.7). Particular rooms may project a sense of warmth. A kitchen fosters an atmosphere of 'friendliness and cheerfulness' (2.13.9), or a living-room can feel comfortable and safe:

> I have vivid memories of the living room in my grandfather's house. It wasn't fancy or new, but all old things worn and well used. It had a quality of warmness. There was a stuffed deer's head over the mantle, and I remember once lying on the rug by the fireplace looking up at it. I remember feeling warm and happy, snug and secure (2.7.6).

Use is one quality prerequisite for warmth; a warm home or room will not be one that is unused or used only infrequently. Instead, frigidity and emptiness permeate these places. 'When you're in a house that hasn't been lived in for a long time there's a feeling of coldness,' said one group member (2.7.1). 'There's a definite feeling, a lack of energy in a place where people have not been for a long time,' explained another group member. 'It feels like a ghost town' (2.7.3). 'You walk in and the room feels so cold,' said another group member describing an infrequently used dining-room in her parents' home (2.7.5). Presence of people and interpersonal harmony are integrally linked with use. Atmosphere may change when friends move into a home that before had been shared with acquaintances:

> When we moved into our apartment, some subletters were living with us, but they weren't really friends. They were just helping to pay the rent. When you went home it felt like an apartment, it didn't feel too much like home. As soon as our friends moved in the place changed. It's nice now. Even when I return and nobody's home, there's a good aura about the place. We all get along fine, we eat dinner together every night, and it's just like home. You look forward to being together, eating together. It seems like we've developed a family feeling (2.10.1).

Care is associated with places of warmth: the person feels concern for the home and keeps it ordered and in good repair. The place, in turn, radiates a sense of tidiness and quiet beauty. 'The place looked nice,' said the group member of the friend's house which he vividly remembered. 'Someone had taken pains with the place' (2.7.4). 'It was decorated in a light blue and was clean and ordered and cared for,' said the group member visiting the home which made him wish to be a child living there (2.7.7). Cleaning and fixing a place that originally feels 'cold and unused' may help the dweller come to like it and feel at home there (2.7.2).

Warmth, unlike other aspects of at-homeness, is a less tangible quality and not present in all homes. Warmth, however, is not insignificant. It sustains an atmosphere of cheeriness and companionship which enhances the sense of life. Today, many people live alone in homes isolated from any wider interpersonal context. Warmth is less likely to be present and one cannot help wondering what impact its loss has on person and society (Heidegger, 1971).

11 IMPLICATIONS FOR ENVIRONMENTAL THEORY, EDUCATION AND DESIGN

Only if we are capable of dwelling, only then can we build – Heidegger (1971, p.160).

For people in other times and places, at-homeness may involve other dimensions besides those explored here. For the Appalachian mountaineer, the gipsy, the migrant worker or the suburban house-owner the specific nature of at-homeness will vary. No matter what the specific historical and cultural context, however, at-homeness will surely be present in some fashion, whether it involves a small parcel of land on which a small log cabin sits, a series of sites at which a nomad stops on his annual migratory travels, or the motel room of a travelling salesman.

The five aspects of at-homeness discussed here touch different parts of the person, have varying spatial manifestations, and lead to different experiential consequences. Table 11.1 summarises these variations. Rootedness touches largely the bodily part of the person, providing both spatial and temporal orientation. Rootedness is strong within the home and extends outward to the places and paths which the person uses actively. Rootedness supports order of movement and continuity in time. Present and future behaviours can happen as they have in the past in the space where a person is rooted. He need not plan and figure out day-to-day activities and events; rootedness guarantees their automatic unfolding.

Rootedness has been ignored in much research on territoriality, though some studies of animals have noted its significance (e.g. Von Uexküll, 1957; Lorenz, 1966). Leyhausen (1970, p.184), making reference to Hediger (1949), explains that mammals establish a territory consisting of points of interest – i.e. first-order homes, second-order homes, places for feeding, rubbing, sunbathing, etc. – 'connected by an elaborate network of paths, along which the territory owner travels according to a more or less strict daily, seasonal, or otherwise determined routine'. Applying this pattern to domestic cats, he finds they have such a territory. He can not definitely conclude however, that they have a definite routine as Hediger had maintained. Leyhausen writes:

Table 11.1: Aspects of At-Homeness

Aspect	Experiential Stratum	Manifestation in Space and Place	Consequences
Rootedness	bodily, centres the person spatially	concentrated in places, paths, points of use; undeveloped in unused portions	intelligent body-subject; spatial order and temporal continuity; minimal chance of being or becoming lost; taken-for-grantedness in terms of orientation, routines and places for things
Appropriation	largely emotional; attachment to place (positive), sense of threat (negative), sentiment	concentric and generally strongest at centre; intensity in proportion to use and attachment; applies to centres, paths, places for things and things themselves	provides person a place of ownness and order in a wider world that is public, often chaotic
Regeneration	bodily, emotional, cognitive; renewal and repair	generally happening within the home, but also associated with other places that have restorative powers — e.g. a path where one takes his daily constitutional	restores both body and spirit; repose and sleep
At-easeness	bodily, emotional, cognitive	usually strongest at home but possible in other places where the person feels comfortable and relaxed	relaxation, looseness, contemplation, freedom 'to be'
Warmth	precognitive; felt most by the person's bodily, emotional parts	most common to interior spaces — e.g. rooms and houses; related to tidiness, decoration, interpersonal harmony	cheerfulness, contentment, sense of camaraderie and nurturing

> In their daily routine, the animals [neighbouring cats] avoid direct encounters, and even cats sharing a home keep separate in the field. According to Hediger, species achieve this by following a rather definite timetable, scheduled like a railway timetable so as to make collisions unlikely. Wolff's and my observations have so far failed to produce any positive evidence that the daily routine of domestic cats is subject to such a definite schedule. Where there is a strong tendency towards being in a certain place at the same time every day, this is usually due to human influence, for example, feeding time (ibid., p.186).

Rootedness is ignored in most studies of human territoriality or reduced to observable behaviours usually called 'activity-space patterns' which are represented in graphic or statistical form (e.g. Boal, 1969, 1971; Lee, 1970; Buttimer, 1972). Rootedness supplements territoriality's emphasis on space as a function of aggression and opens investigation to the role of habitual, precognitive behaviours in joining animal or person with lived-space.

Appropriation incorporates the emphasis in territoriality on protection and defence but extends this affective bond to include positive attachment. Most work in human territoriality has focused exclusively on negative emotions in regard to place, thus Suttles (1968) emphasises the forces of neighbourhood separation, defence and exclusiveness in his study of the socio-spatial dynamics in Chicago's low-income Adams area neighbourhood. Similarly, Newman (1973) falls back heavily on the aggression-defence impulse in his theory of 'defensible space'.

Appropriation lies in the emotional part of the person and sometimes involves a mechanism of defence. The error of territoriality has been the reduction of spatial experience to *only* spheres of influence and control. The aggression impulse is but one part of appropriation. As a complete whole, appropriation involves both negative *and* positive emotions expressed by a myriad of such feelings as attachment, protection, nostalgia and home-sickness (Relph, 1976b, pp.33-43). Tuan (1974) calls the positive emotions for place *topophilia*, which he defines as 'all of the human being's affective ties with the material environment' (p.93), fostered by such diverse experiences as aesthetic appreciation, physical contact, health, patriotism, familiarity and attachment (pp.92-102). Relph (1976a) has developed the term *topophobia* to describe yet another aspect of emotion and place – 'all experiences of spaces, places, and landscapes which are in any way distasteful or induce anxiety and depression' (27).

Appropriation can be visualised spatially as an invisible atmospheric surface whose height at a particular place is directly related to the attachment that the person feels for that place coupled with his need for its appropriation. Generally, this surface is high around a person's favourite possessions, his home, and centres to which he feels close. Typically, its height descends for places further from dwelling-place, though points of attachment far beyond the home – for example, a favourite pub, special swimming place, community where one grew up – may modify the general pattern.

At-easeness can be represented by the same tent-like surface. High points are places where the person is more or less relaxed and free. At-easeness is generally strongest within the physical structure of home, but extends to other places where the person feels comfortable. Territoriality studies have largely ignored at-easeness, probably because its behavioural manifestations are not as readily observable as more visible territorial behaviours like aggression and defence. For the same reason, warmth has been ignored by territoriality research. This quality of at-homeness is most limited spatially, manifesting at special places – generally interiors – which extend an atmosphere of cheeriness and support.

Regeneration points to the restorative powers of place and is best represented graphically as points in space where the person can find physical and psychological rest. These points are probably few for most people. Students in territoriality have sometimes mentioned the significance of place in sustaining regeneration (e.g. Carpenter, 1958) but have not considered the relationship in detail. A study of regeneration in terms of routine and ritual would be one interesting probe. Who sleeps how long, when, where, and how often? What going-to-bed and waking rituals are there for which animals, for which people?

The Development of At-Homeness

The key to identity of place, argues Relph (1976b, p.49) is *insideness* – the degree to which a person belongs to and associates himself with a place. The person who feels *inside* a place is here rather than there, safe rather than threatened, enclosed rather than exposed. The more profoundly inside a place the person feels, Relph explains, the stronger will be his or her identity with that place. Further, Relph suggests that the dualism between insideness and its opposite, *outsideness*, is a fundamental dialectic of environmental experience and behaviour: for different people, different places take on different degrees of insideness and outsideness.

The development of at-homeness is usefully viewed in terms of Relph's inside-outside designations (ibid., pp.49-55). Places of most profound at-homeness generally reflect *existential insideness* – the situation of unselfconscious immersion in place described in Chapter 6 in relation to place ballets. Existential insideness, says Relph (ibid., p.55), is the experience of place 'that most people experience when they are at home and in their own town or region'. It involves fully the five components of at-homeness: the person is bodily and emotionally immersed in place; life holds continuity and regularity and its mundane aspects, at least, are taken for granted and rarely reflected upon.

How does at-homeness develop? Having no home and not being at home are reflected in two of Relph's modes of outsideness: *existential outsideness*, where the person is homeless and alienated from place and people (ibid., p.51); and *objective outsideness*, where the person intentionally adopts a dispassionate attitude toward place in order that it can be studied selectively in terms of specific locational or activity attributes (ibid., pp.51-2). People in these modes of experience are separate from place and feel no attachment. They are entities severed experientially from the geographical world in which they find themselves. They are not at home but observers, tourists, people alone or 'out of place'.

Becoming at home is reflected in Relph's modes of insideness. In *behavioural insideness* the person begins to notice place – he deliberately looks for aspects of place that make it different from other places. A uniform environment begins to reveal pattern and uniqueness; the person begins to feel inside and at home (ibid., pp.53-4). As these feelings strengthen, the person may experience *empathetic insideness*, which requires a 'willingness to be open to significances of place, to feel it, to know and respect its symbols' (ibid., p.54). The person grows closer to place and feels attachment and concern. Eventually, but not necessarily, the person may choose to make this place his home; he experiences *existential insideness* and a 'deep and complete identity with place' (ibid., p.55).

At-Homeness Past and Present

At-homeness in the past sprang from existential insideness. People born in place lived there the rest of their lives. They became at home in that place. At-homeness was experienced unselfconsciously and never questioned unless suddenly upset by natural disaster or social upheaval. Clearly there were exceptions – travellers, outcasts, the well-to-do. For most people, however, birth place was the only place. It was home, no matter how wretched and unfair conditions might seem to the outsider.

The poet and agricultural writer Wendell Berry (1977) argues that in this past mode of at-homeness, people lived in intimate contact with place and natural environment. Aspects of living were continuous, and each place was profoundly unique:

> Once, some farmers, particularly in Europe, lived in their barns – and so were both at work and at home. Work and rest, work and pleasure, were continuous with each other, often not distinct from each other at all. Once, shopkeepers lived in, above, or behind their shops. . . Once, households were producers and processors of food, centers of their own maintenance, adornment, and repair, places of instruction and amusement. People were born in these houses and lived and worked and died in them. Such houses were not generalizations. Similar to each other in materials and design as they might have been, they nevertheless looked and felt and smelled different from each other because they were articulations of particular responses to their places and circumstances (ibid., p.53).

Today, in an era of mobility and mass communications, people easily transcend physical space and readily compare and switch places. At-homeness is no longer a certainty; it must be re-established each time a person moves. Many people are footloose and feel no attachment to place. At the same time, technology and mass culture destroy the uniqueness of places and promote global homogenisation. The result is the placelessness of which Relph (1976b) speaks.

Berry (1977) goes on to argue that the modern mode of at-homeness is responsible for placelessness and community fragmentation. The home is no longer grounded in place, he says, and when people do not live where they conduct other aspects of their lives such as working and recreating 'they do not feel the effects of what they do' (ibid., p.52). People, for example, who dig strip-mines, who build expressways, who clear-cut forests 'do not live where their senses will be offended or their homes or livelihoods or lives immediately threatened by the consequences' (ibid., p.52). Home is no longer a kindly response to a specific physical and social milieu; rather, it is

> a generalization, a product of factory and fashion, an everyplace or a noplace. Modern houses, like airports, are extensions of each other; they do not vary much from one place to another. A person standing in a modern room anywhere might imagine himself anywhere else – much as he could if he shut his eyes. The modern house is not a response to its place, but rather to the affluence and social status of its owner. . . His home is the emblem of his status, but it is not the center of his interest or of his consciousness. The history of our time has been to a considerable extent the movement of the center of consciousness away from home (ibid., 52-3).

At-Homeness and Dwelling

Martin Heidegger (1962, 1971), the influential German philosopher and phenomenologist, has perhaps discussed in most detail the situation of modern homes and at-homeness. Heidegger believes that we are forgetting how to dwell and by that forgetting, we also forget how to be at home and how to build homes. *Dwelling*, says Heidegger (1971), is the process through which man makes his place of existence a home and comes into harmony with the *Fourfold* – the earth, sky, gods and himself. To dwell is 'to be on the earth as a mortal', which in turn means 'to cherish and protect, to preserve and care for' the place and community where one chooses to live (ibid., p.147). Dwelling, explains Buttimer (1976, p.277), placing the notion in modern context, 'means to live in a manner which is attuned to the rhythms of nature, to see one's life anchored in human history and directed to a future, to build a home which is the everyday symbol of a dialogue with one's ecological and social milieu'.

Dwelling incorporates at-homeness and extends to other themes: the quality of environment and place that sustains or hinders a home, the manner by which people might better treat the earth and land, the responsibility of the individual for himself and others. While at-homeness focuses largely on the context of person and needs of people, dwelling, reaching further, has ecological significance and joins the individual person with earth, biosphere, communal and spiritual milieu. Dwelling points towards man's role of *caretaker*: of the natural world, of other people, of himself, of kindness, morality and understanding (Grange, 1977).

The key to dwelling and caretaking, Heidegger (1971, p.149) says, is *sparing and preserving* – the kindly concern for land, things and people as they are and as they can become. Sparing and preserving foster a 'free sphere that safeguards each thing in its nature' (ibid., p.149). The result, Relph explains,

> is places which evolve, and have an organic quality, which have what Heidegger calls the character of 'sparing' – the tolerance of something for itself without trying to change or control it – places which are evidence of care and concern for the earth and for other men. Such spaces and places are full with meaning; they have an order and a sense that can be experienced directly, yet which is infinitely variable (1976b, p.18).

Dwelling and Building

'To build is in itself already to dwell,' Heidegger goes on to say (ibid., p.146), striking the essence of the value of dwelling for environmental education and design. Modern men and women, argues Heidegger, have forgotten how to dwell and can no longer build. 'What does it mean to live in places fully?' Until we answer this question, Heidegger says, we can not expect that the planned spaces we construct will be successful human or ecological environments.

Dwelling, if we trust Heidegger, is more than attractive buildings and surroundings, or needs defined by physical criteria – amount of floor space, lighting, or whatever. Rather, dwelling involves less tangible qualities and processes – caring for the place where one lives, feeling at home in and part of that place. Until we as social scientists and planners begin to understand the day-to-day environment in terms of dwelling, Heidegger argues, we will continue to create locations that are lifeless and empty spaces rather than lived-in places fostering a sense of vitality and community.

At the same time, Heidegger, like Berry, believes that people exist in place but no longer dwell there. Residents in public housing who vandalise, even destroy their living environment; suburbanites who attack a 'they' named responsible for a badly constructed sewer system or mismanaged landfill; corporate executives who become outraged at environmental regulations – people like these, argues Heidegger, forget their responsibility to earth and the larger human community. They move towards a state of homelessness and interpersonal alienation. Using housing as an example, Heidegger writes:

> On all sides we hear talk about the housing shortage, and with good reason. Nor is there just talk; there is action too. We try to fill the need by providing houses, by promoting the building of houses, planning the whole architectural enterprise. However hard and bitter,

> however hampering and threatening the lack of houses remains, the *real plight of dwelling* does not lie merely in a lack of houses. The real plight of dwelling is indeed older than the world wars with their destruction, older also than the increase of the earth's population and the condition of the industrial workers. The real dwelling plight lies in this, that mortals ever search anew for the nature of dwelling, that they *must ever learn to dwell* (ibid., p.161, italics in original).

Learning to Dwell

Learning to dwell happens in two ways: discovering the importance of dwelling in our own and others' lives; designing physical environments that sustain and enhance dwelling.

The environmental experience groups provide evidence of the first way. Through exploring experiences of place and space, some group members came to recognise the importance of home and at-homeness (Appendix B). There was growing understanding of the numerous experiential links binding person with place. Consider, for example, the following report made by a group member the summer after the group experience; she has become sensitive to the value of home and sees its importance in day-to-day living:

> One thing I noticed over the summer was the way the house I was living in became a home. There was a sense of centre that survived moodiness and restlessness. How important that was! All sorts of little things helped – someone lent me sheets, so I didn't have to sleep in a sleeping bag any more, and I moved the bed back to the bedroom – it was in the living room when I came – so that there was a living space and sleeping space. Having this home, as temporary as it was, ended up to be one of the best things about the summer. Real delight in cleaning things up, putting things in their places, having a place that was mine (commentary 3).

If education about dwelling is valuable, finding ways to construct physical environments that support rather than stymie dwelling is also important. One returns again to community and place ballet. In one sense, it is fair to say that people's striving for dwelling and at-homeness is the experiential foundation of community. Community establishes a familiarity and comfortableness with people and environment outside the immediate home. Community and place in the past were bounded areally; human movements were restricted by the distance a person could walk or ride by relatively slow means of transportation. Residence, business, work and recreation came together in space and time. Place ballet and caring for place were readily possible. Modern time-space routines, in contrast, are often isolated units which rarely fuse in a wider place-space whole. Activities are segmented; the potential interpersonal dynamism that might result if they did fuse is lost. 'Geography', says Berry (1977, p.53) about modern man, 'is defined for him by his house, his office, his commuting route, and the interiors of shopping centers, restaurants, and places of amusement – which is to say that his geography is artificial; he could be anywhere and usually is.'

Clearly, advances in technology allow people to overcome distances physically and make their lived-space the changeable patchwork which Berry describes. Experientially, however, this physical separation of places has a part in destroying community. If they are close spatially, people are more likely to develop interpersonal ties and to care for the space they share. Community, as Slater (1970, p.5) reminds us, is 'a total and visible collective entity'. Alienation from space weakens this visible totality and destroys concern for specific places and environments.

Dwelling, At-Homeness and Place Ballet

In asking, then, how physical design can foster dwelling and at-homeness, one returns to the notion of place ballet. First, place ballet *roots* the person in his own time-space routine as at the same time it *roots* the totality of participants in an overall time-space pattern. Second, in that it automatically 'assigns' specific people to a specific area, place ballet *appropriates* space and thereby fosters interpersonal familiarity and trust. At the same time, this acquaintance and sociability may generate *at-easeness*, and in some environments *warmth*. Place ballet fosters a sense of place which provides participants with spatial order and identification as it protects them from the intrusion of uninvited people and events from the world at large.

If place ballet promotes at-homeness and dwelling, there still remain two important questions: what specific elements of physical design promote or hinder place ballet? What specific place ballets are needed by which places?

The first question I momentarily set aside but return to in Part Five. The second question presently has no clear answer. In urban environments, if the argument of Jacobs (1961) is trusted, place ballets centre on neighbourhoods, streets, their shops and other establishments. The need is to promote situations where place ballets can be strengthened, revived or begun, and their dynamism spill out into less lively surroundings. In rural areas, suggests Berry (1977, pp.218-22), the need is a return to the individual landowner with his family and care for the land. Each farm would promote a family place ballet that mingles with a larger, regional time-space whole. This interaction, says Berry, would re-establish a face-to-face relationship between ruralites and urbanites.

In today's technological and mobile world, a return to human scale may seem impossible. Webber says of the modern American:

> The communities with which he associates and to which he belongs are no longer only the communities of place to which his ancestors were restricted; Americans are becoming more closely tied to various interest communities than to place communities, whether the interest be based on occupational activities, leisure pastimes, social relationships, or intellectual pursuits (1970, p.536).

On the other hand, social and spiritual discontent, as well as economic, ecological and energy problems rumble beneath the mainstream and suggest uncomfortably that our world is not secure. Countries throughout the world are becoming increasingly mobile. One out of every three American families changes place of residence every three years; Canadian and European trends begin to reflect a similar pattern. People no longer feel a sense of 'belonging'. There is less of an effort to become part of place and community. People are less likely to feel responsible for their neighbours because they don't know them.

In terms of experience, which is the main test here, it does seem that we are unavoidably grounded in place and space and this is partially because we are bodily and emotional beings. If we were gaseous creatures or finer energies that could fly instantly from place to place, if we did not develop emotional attachments, our day-to-day geography would be unimaginably different. The role of place and space might be insignificant.

The decision, ultimately, as to what we are lies with the individual person who can decide that man is infinitely adaptable or, alternately, that he is limited by various inescapable qualities that include body, feelings – the need to dwell. Experientially, at least, people can not do everything. The best solution, perhaps, is to recognise existential limitations and to become free around them – not in spite of them.

Part Four

ENCOUNTER WITH THE GEOGRAPHICAL WORLD

The earth is all before me.
With a heart Joyous, nor scared
at its own liberty, I look about
– William Wordsworth (1936, p.495)

12 PERCEPTION AND A CONTINUUM OF AWARENESS

> My wife is getting blind; on the whole she is glad of it. There is nothing worth seeing. She says she hopes she will also become deaf; for there is nothing worth hearing – Strindberg (cited in de Grazia 1972, p.473).

I round a curve on the road and suddenly notice the brilliant autumn foliage ahead; I enter a corner grocery store and observe that its doors have recently been repainted; I wait for a bus and watch the children skating on the pond across the street. In each of these experiences, a part of my awareness has touched and been touched by an aspect of the geographical world; a strand of attention is present between me and the trees, the building, the pond as a place of activity. These moments are all representative of *encounter – any situation of attentive contact between the person and the world at hand.*[1]

Unlike movement and rest, which are both clearly observable phenomena in the natural world, encounter is less obvious because it deals with inner situation as well as external entity or event. The student can easily conclude through observation that a rock is resting or that a cat may have a daily time-space routine. To observe encounter in other forms of life besides man, however, is more difficult because there is no easy way to detail inner situation as it meets external world.[2] The possibility that other entities in nature besides people encounter their surrounding world is frequently met with scepticism by contemporary scientists and the general public. Primitive and traditional cultures, however, often held that animals, trees, rocks and places possessed a certain awareness of themselves and their world (Gutkind, 1956; Eliade, 1957; Searles, 1960; White, 1967; Relph, 1976b), and recent studies provide some supporting evidence for such beliefs (Schwenk, 1961; Backster, 1968; Roszak, 1969 and 1973). As conventional science accepts the existence of finer, less readily identified phenomena, perhaps the possibility of encounter for other entities besides man will be explored and it will not seem so strange to speak of the 'sentience' of a rock, or the 'spirit' of place (Banse, 1969; Durrell, 1969; Roszak, 1973; Seamon 1978a).

Encounter has generally been described in terms of *perception* in traditional philosophy and psychology. Perception refers to 'the way an observer relates to his environment' (Murch, 1973, cited in Ittelson *et al.*, 1974, p.103). Perception, says Allport,

> has something to do with our awareness of the objects or conditions about us. It is dependent to a large extent upon the impressions these objects make upon our senses. It is the way things look to us, or the way they sound, feel, taste, or smell. But perception also involves, to some degree, an understanding awareness, a 'meaning' or a 'recognition' of these objects (1955, p.14).

Perception, like spatial behaviour, has most often been described in terms of behaviourist or cognitive theories.[3] Perception, argue the behaviourists, is the process by which stimuli outside the person become signals for the receipt or non-receipt of a reinforcer (Ittelson *et al.*, 1974, p.66): I notice the autumn foliage, for example, because a similar situation in the past has favourably aroused my attention and works as a positive reinforcement for my noticing in the present moment. The environment is a source of information, say the cognitive theorists in contrast. They suggest that I, the observer, take some active part in receiving and structuring the impressions of trees – that is, my perceptual system does not passively absorb their impact as a learned result of past reinforcement, but rather, has an initiating role in my acceptance or rejection of them. This approach to perception has been called the 'information-processing view' (Ittelson *et al.*, ibid., pp.109-13) because the person is likened to a cybernetic device which applies a machine-like, deciding process to the perceptual data. The cognitive approach to perception, unlike its behaviourist counterpart, has had significant impact in behavioural geography. It is the theoretical structure underlying most models of environmental perception in geography (e.g. Downs, 1970) and guides specific empirical research, as, for example, work in landscape assessment and preference (e.g. Zube (ed.), 1976).

Perception, viewed phenomenologically, is 'the *medium of intercourse* between the world that is known and the person who perceives and knows it' (Keen, 1972, p.91, italics in original). Neglect of the experiential aspects of this 'medium of intercourse' is a major weakness of both cognitive and behaviourist theories. They ignore the possibility that moments of perception may vary in quality and intensity. Some students of perception recognise at least in theory that perception is not of uniform impact (e.g. Hirst, 1967, p.80). In practice, however, the varying intensities of perception are usually forgotten. It is studied instead according to categories derived from the five senses (sight, sound, touch, taste, smell) or the external world (colour, form and shape, movement of objects, illusions, space, time, or some similar theme).[4]

To explore moments of perception happening as experiences is the aim here. I seek to resurrect the experiential integrity of perception before it has been reduced and flattened by a particular psychological or philosophical theory. People, it will be found, encounter their world in a variety of ways. A moment of encounter is integrally related to other aspects of the moment, including mood, energy level, past experience and knowledge.

At different moments, the person pays more or less attention to the world at hand. At times, he is intensely aware of the environment and may even feel that he is in perceptual union with it. After Krawetz (1975), I identify encounters of this sort by the term *tendency towards mergence* because there is a break in the boundary between person (self) and world (non-self); in figurative terms, the person *merges* with his environment. At other times, the person is very much oblivious to the world at hand and gives it no notice. I identify the term *tendency towards separateness* with these kinds of encounters because the person is directing his attention inwardly, and is *separate* (in terms of awareness) from the world at hand. Even in situations of extreme separateness, however, the preconscious perceptual abilities of body-subject are at work, guarding the person from any unexpected dangers that the environment might impose, and assisting with any gestures or movements that the person is required to make even as his more conscious attention is directed elsewhere.

Encounter is not one kind of experience but several, whose sum may perhaps be best described as an *awareness continuum* that incorporates on one side encounters tending toward mergence, and on the other encounters tending toward separateness (Figure 12.1). Exploring the nature of encounter leads to a better understanding of how human beings attentively meet the places, spaces and landscape which are their surrounds. This knowledge has considerable import for research on specific environmental perceptions as well as for environmental education. In addition, it can be asked how various kinds of encounters relate to a situation of at-homeness.

Figure 12.1: Modes of Encounter: An Awareness Continuum (for waking states)

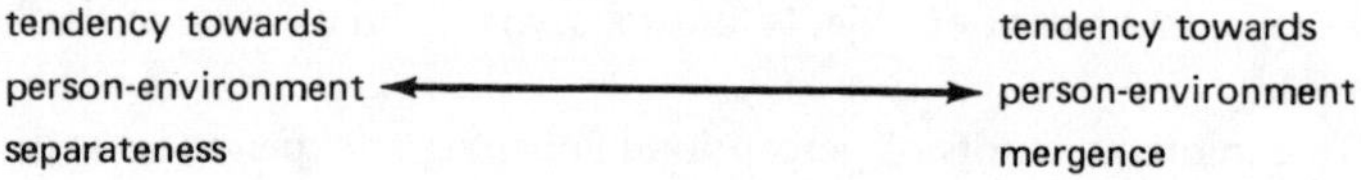

Notes

1. Though my concern is with the geographical world, I recognise at the start that encounter extends to all other aspects of the world – its interpersonal, social, cultural, economic, historical, spiritual dimensions. The person watching the pond, for example, is participating in interpersonal and social encounter, since he is looking at people and their interactions, as well as absorbing the atmosphere of that pond as a place. What is needed is a phenomenology of encounter which investigates all encounter – not just with the geographical world. The illustrations of encounter in this book involve attentive contact with aspects of the physical non-human environment, and also include examples of encounter with people, things, places and events housed within these places.

2. Even for human beings, scientific devices such as lie-detectors can measure recordable dimensions of inner situation such as pulse rate or muscle tension, but cannot provide an account of that situation as it is in experiential terms.

3. A good philosophical introduction to perception is Hirst, 1967. Merleau-Ponty's *Phenomenology of Perception* (1962) is perhaps the most penetrating critique of traditional philosophical and psychological approaches to perception. Ultimately, perception is a key notion in Merleau-Ponty's own philosophical conception of human experience, but his definition is considerably different from empirical (behaviourist) and rationalist (cognitive) interpretations.

As with spatial behaviour, there are not two theories of perception but several. In addition to the major divisions between behaviourist and cognitive theories, there is also the gestalt approach to perception which marks a significant third interpretation. See Allport (1955) for an overview of these various approaches.

4. For examples of these divisions, see Schiffman (1976).

13 FLUCTUATION, OBLIVIOUSNESS AND WATCHING

> It's strange how the world is for me. At times I know I'm not seeing anything. I'm so caught up with my own self inside that the world has no chance to penetrate. Sunday morning, for example, I went for a walk down Main Street. The trees seemed so beautiful and alive. I hadn't seen them in that way in a long time. I was feeling good – in a calm and quiet mood. I happened to meet my old girlfriend who said something that really hurt me. I kept walking but now that walk was completely different. I was full of anger inside and didn't notice a thing. It was like a barrier had been put up. My anger and bad thoughts blocked out the possibility of seeing the way I had a few minutes before – a group member (3.1.2).

Encounter with the world at hand is constantly fluctuating – becoming more or less sharp as the person's attention moves between inner and outer concerns. 'Sometimes', said one group member, 'I'm very close to the world and other times distant and non-alert.' 'Some days I don't notice a thing. . . Other days will seem fresh' (3.1.1). Even in a matter of seconds, degree of awareness to the world can drastically shift; strong contact at one moment may lead to lack of contact next. Note the change in the chapter's opening observation. The group member, satisfied and serene, has penetrated the boundary between self and world: the trees are 'beautiful and alive'. The chance meeting with a former girlfriend, however, erects an inner 'barrier', and the outward encounter is lost.

Awareness of the world continually advances and retreats like the action of waves on shore. Each moment of encounter is unique, of a specific intensity and quality of contact not to be exactly duplicated in other moments. Underneath the flux and variety, however, are a certain few kinds of encounter which can be called *obliviousness, noticing, watching, heightened contact* and *basic contact.* These modes are not exact points of encounter, but rather imprecise benchmarks summarising a particular range of encounters on the wider continuum of awareness described in Chapter 12.

Obliviousness

Obliviousness refers to *any situation in which the experiencer's conscious attention is not in touch with the world outside but directed inwardly* – to thoughts, feelings, imaginings, fantasies, worries or bodily states which have nothing or little to do with the world at hand. Obliviousness implies not a cessation of all attention but only that directed outwardly.

Obliviousness occurs in all variety of contexts. One group member, 'caught up in thought about the week ahead', missed a turnpike exit (3.3.2). A second group member, walking hurriedly, passed by a friend without noticing him (3.2.2). Another group member, lost in thought, suddenly noticed a friend walking in front of her whom she was oblivious to a moment before:

> I was not conscious of his presence, and it surprised me when I noticed him that I hadn't seen him before I did. He was there in front of me several seconds, yet it took a bit of time before I consciously realised he was there (3.2.1).

Obliviousness extends to activities, especially those involving drudgery or repetition. 'When I'm working at my job as dishwasher,' said one group member, 'I rarely pay attention to what I'm doing. It's easier to daydream or think about what I'll do after work' (3.5.2). Sometimes, people forget if they have completed a certain portion of a job. 'In housekeeping it's so easy to go off in a daze,' said one group member who couldn't remember if she had vacuumed one corner of a room and cleaned it again to be sure (3.4.1). People may actually encourage a state of obliviousness as they work. Said one group member, asked why she sang as she did cleaning chores, 'Well, you've got to do something to take your attention off washing up' (3.5.1).

Although obliviousness is sometimes associated with positive inner states (3.3.5); it more often occurs in situations of sickness, hurry or negativity. 'I know when I'm not feeling well physically,' said one group member, 'I just look at the ground and try to get where I have to go. There's no energy left to notice things' (3.3.4). Times of hectic activity may close the person to the world beyond. 'I'm always running around,' said one group member, 'thinking about what I should do next. . . In times like these, I don't notice many things around me' (3.3.1). Anger is especially effective in blocking out awareness, as the person meeting his girlfriend indicates. Sometimes the anger may sever the person entirely from the situation at hand and he may make a mistake (3.7.2).

Negativity and tiredness make it extremely difficult to attend to the world at hand, even if one becomes aware of his obliviousness. Said one person, walking home and tired from a long bus ride:

> When I got off the bus, I noticed how miserable I felt, tired and hungry. I just wanted to stop travelling. About half the walk was over before I noticed that everything was passing me by. I wasn't making connection at all. I saw my situation and tried to get more into the environment. But five seconds later I had drifted off again. I was so 'tuned out' that I went trudging across some grass that was a shortcut. I didn't notice it until the deed was done and then had a good laugh at myself (3.15.3).

Obliviousness is a range of experiences in which the person is more or less unaware of the world at hand. On the awareness continuum, obliviousness is associated with the tendency towards separateness; it is placed on the extreme left as in Figure 13.1.

Figure 13.1: The Place of Obliviousness on the Awareness Continuum

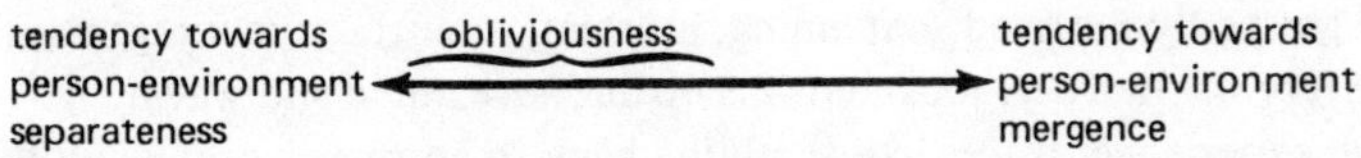

Watching

Watching is *a situation in which the person looks out attentively upon some aspect of the world for an extended period of time.* Watching is of different types and intensities, ranging from a sporadic, weakly directed variety to strong emotional and bodily involvement (Figure 13.2).

Watching of weakest intensity occurs in situations where the person

Figure 13.2: The Place of Watching on the Awareness Continuum

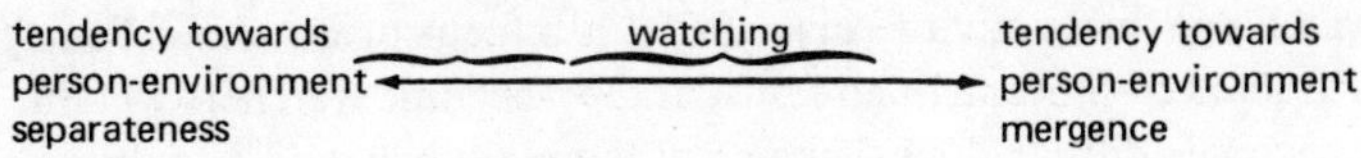

is only peripherally concerned with the world at hand. His attention wanders back and forth between inner and outer concerns. Consider a group member relaxing on a dormitory lawn:

> I was sitting on the lawn in front of Wright Hall Tuesday afternoon watching people, seeing who was going where. I wasn't watching anything or anyone in particular – just looking. It's relaxing. I must have sat there an hour or more, taking in the atmosphere. I'm not saying I was watching the scene the whole time. Sometimes I'd be 'into myself', thinking about things or worrying about school work I should be doing. It was a mixture – lost in thoughts for a while, then noticing something, on and on (3.8.1).

At other times, watching may be more intense as the interest, beauty or excitement of the scene draws the person's attention there and holds it. One group member, sitting on a park bench, watched ducks on the pond before her. 'It was like watching a movie,' she explained (3.8.2). Another group member, spectating at an automobile race, described a watching considerably more intense. His body and his attention were involved in the spectacle at hand:

> There were three racers scrambling for the lead, and no one had a clear edge. A friend owns one of the cars, and I was cheering for him. I got really involved – standing, jumping, shouting encouragement. Everyone in the grandstand was up and screaming and waving. It was an experience. It was like tumbling back to another world when the race ended (3.8.3).

Watching requires activity and movement. People do not normally watch things and places that are inactive. Busy pavements, noisy ducks, racing automobiles foster noticing in the above examples. Who would watch empty streets, an unused pond, a dormant speedway? Such situations are neutral backdrops and normally do not capture the experiencer's attention.

People often generate the activity and movement responsible for watching. 'Nobody', says Jacobs (1961, p.35), 'enjoys sitting on a stoop or looking out of a window at an empty street. Almost nobody does such a thing. Large numbers of people entertain themselves, off and on, by watching street activity.' Place ballet is a focus of watching. People are attracted to movement and bustle; they become watchers who in turn become additional participants in the place ballet. 'The sight of people attracts still other people', says Jacobs (ibid., p.37). She goes on to describe a stretch of upper Broadway in New York City along which much watching occurs:

> People's love of watching activity and other people is constantly evident in cities everywhere. This trait reaches an almost ludicrous extreme in upper Broadway in New York, where the street is divided by a narrow central mall, right in the middle of traffic. At the cross-street intersections of this long north-south mall, benches have been placed behind big concrete buffers and on any day when the weather is even barely tolerable these benches are filled with people at block after block, watching the pedestrians who cross the mall in front of them, watching the traffic, watching the people on the busy sidewalks, watching each other.

Watching adds to the dynamism of place ballet. Its watchers, especially if they participate regularly, work as unknowing caretakers. They regularly watch their place, not for remuneration but because it is the taken-for-granted thing to do. Perhaps they offer assistance to strangers or notify the police when trouble occurs. 'Thousands upon thousands of people. . .', says Jacobs (ibid., p.38), 'casually take care of the streets. They notice strangers. They observe everything going on.'

Regular watchers are familiar and comfortable with their place – they are *at home* there. They feel responsibility for place and work to protect and care for it. Such concern can arise only where watcher-participants know each other and the regularities of place. These people can count on mutual support in times of trouble and are willing to exert proprietorship when the place is threatened in some way.

Watching establishes an extended span of attention between person and place. To watch is to pay attention at length to the world at hand – to have one's interest occupied as mutually the world receives that interest. Normally, watching does not arise because of conscious planning on the part of the watcher, but because the world automatically attracts and holds the watcher's interest in some way. Watching is essential for places filled with human activity because it demands and sustains behaviours which are publicly proper for the particular place. One means of fostering watching, therefore, is to foster place ballet. This possibility is considered further in Chapter 19.

14 NOTICING AND HEIGHTENED CONTACT

> We walked by an alley that I'd not noticed before. . . It was something I had never seen, yet I had passed that place many times. I don't know what caused me to notice it – a group member (3.10.1).

Noticing is sudden. *A thing from which we were insulated a moment before flashes to our attention.* Noticing is *self-grounded* or *world-grounded.* Personal knowledge and past experience trigger the former; some striking characteristic of the world sparks the latter.

Incongruity, surprise, contrast and attractiveness (or its opposite, unattractiveness) are all characteristics that activate world-grounded noticing. One group member, travelling through a wilderness area, unexpectedly came upon a 'huge man-made machine' in the middle of a stream. 'It stopped us,' she explained. 'It seemed out of place and took our attention' (3.11.1). Another group member, riding on a turnpike, became aware suddenly that he had arrived in Delaware because the texture and colour of the road changed (3.11.4). A third group member described the sudden impact that a sunny field of freshly harvested pumpkins had (3.12.3), while yet another group member took notice of a bank building because it was 'round and hideously designed' (3.12.3).

The world grabbing one's awareness is the mark of world-grounded noticing. 'Took our attention' in the first report encapsulates well the experiencer's passive role: attention is immediately 'taken up' by the machine because it contrasts so completely with the taken-for-granted wilderness through which the group member has travelled for the past several days. She has little choice not to notice.

Person-grounded noticing gives the individual a more active role in awareness. It generally involves things about which the person wants to know more. *Wish* and *need* to understand provide a context of interest out of which noticing may occur, but do not guarantee noticing. The instant of awareness, like world-grounded noticing, is spontaneous and unexpected.

Consider the following account dealing with coloured shadows – a common phenomenon in the everyday environment, but one which most people usually don't notice.[1] The report indicates two variations of person-grounded noticing: one where the group member looks actively and then notices; the other where noticing occurs spontaneously. In both cases, the actual moment of seeing is sudden and uncontrolled:

> I never used to notice coloured shadows – in fact, I never knew they existed. Yet because I took a course that spent a lot of time studying them, I've become aware of coloured shadows and look for them when I think of them. I notice them often now, especially in the streets at night. The best thing is that the more I notice coloured shadows, the more I look for them. At first, I was aware of them only rarely, but now I notice them quite often. It's not that I walk down the street saying to myself, 'Okay, it's time for you to be conscious of coloured shadows.' Rather, the thought of them will suddenly pop into mind, something in me will look, and maybe I spot one. Or sometimes, I'll be walking along in a daze, and I suddenly notice one. They jump out at me – I don't make any active effort to see them. It's as if they show themselves to me and I don't do a thing but respond to them (3.13.2).

Sometimes, says the group member, noticing occurs because he thinks about coloured shadows and actively looks outward. This process is rapid, unfolding instantaneously in one smooth flow. 'The thought. . .', he says, 'will suddenly pop into mind, something in me will look, and maybe I'll spot one.' At other times the noticing is completely unexpected: 'I'll be walking along in a daze and suddenly I'll notice one.'

The unexpectedness of noticing follows no clear pattern. Different people notice different aspects of the same environment: a thing noticed one time may not be noticed at other times (3.9.1, 3.10.1). Noticing, suggests the above observation, leads to more noticing. 'The best thing', says the group member, 'is that the more I notice coloured shadows, the more I look for them.' Training and interest provide some control of noticing. Ultimately, however, it is unpredictable, happening in moments for which the person can not plan.

Inner state is closely related to noticing. Positive moods enhance noticing. One group member, pleased with the photographs he had done, spent an entire afternoon taking more pictures. 'I was noticing more than I usually do,' he said, 'and it had something to do with the fact that the photographs had come out so well' (3.15.2). Another group member, feeling exuberant after a rewarding discussion, found herself actively engaged with the park through which she walked, noticing a great deal:

> It seemed that I was noticing a lot because I was feeling so good. The ducks on the pond, the colours reflecting on the water, the trees – I was very much aware of them. I was noticing more around me than I usually do (3.15.1).

Negative states may foster noticing, but of a sort that highlights the unpleasant and disturbing parts of the world. One group member noticed troublesome smells when she was negative (3.15.4), while another group member became aware of other people's unpleasant qualities (3.15.5). A third group member, describing a shopping trip, pointed out in detail the impact on noticing that negativity can have:

> Last Thursday afternoon I went grocery shopping. Everything seemed to go wrong. The store was out of some things I needed, the cashier wouldn't accept my foodstamps because I didn't have my identification card. I was angry. Everywhere I looked I saw another thing which showed in more detail what a mess the world was in. I remember noticing the four big metal pillars that hold up a huge electric sign for the supermarket where I had been shopping. There was this quick flash of annoyance in me that was saying, 'What a waste of resources – all that people in this crazy country know how to do is waste.' I was startled to see myself upset by so little a thing as a sign, but I couldn't shake my negativity. I went home and went to bed (3.15.6).

Noticing makes the unnoticed world known, without required participation or desire of the noticer. 'Sudden and unmeditated' are the essential characteristics of noticing. If it weren't an integral mode of human encounter, we would be required to direct each attentive contact with the world actively, just as, if there were no body-subject, we would need to direct each gesture and movement cognitively. Noticing brings the world directly to our awareness. Self and non-self come momentarily together through the power of attention.

Noticing involves a direct, attentive meeting between person and world, and therefore tends more towards an experience of mergence than watching. On the awareness continuum I place noticing as in Figure 14.1. Realising that some watching may be more intense than noticing, I overlap the two, just as I did watching and obliviousness.

Figure 14.1: The Place of Noticing on the Awareness Continuum

Heightened Contact

In *heightened contact, the person feels a serenity of mood and vividness of presence; his awareness of himself is heightened, and at the same time, the external world seems more real.* Consider the following accounts:

> I was sitting on my usual bench, facing the two brown houses and pine trees behind the court where I usually play tennis. As I was gazing at the pine trees, the sun went behind the clouds for a moment and some noisy birds flew over the houses. Suddenly, I felt very still but shivery inside. I felt quiet, I felt as if all the world was at peace. I felt warm towards everything (3.16.1).

> One day this past summer I was driving across the Verrazano Bridge. All of a sudden I felt very high emotionally and in harmony with everything around me. The bridge stood out as a strong, all-consuming structure, yet at the same time, I felt connected to the bridge in some kind of spiritual way. The moment lasted as long as I was on the bridge. It was vivid and I clearly remember it (3.16.2).

> On Wednesday, I visited a museum to do some research on the Shakers. I had driven quite far on the forested backroads, but the whole trip I hadn't really noticed anything. I was caught up in my thoughts. When I reached the museum I drove the car down the long driveway and parked it in the lot which overlooks a valley. As I got out, I had this strong experience. I felt a rush of warm spring air on my face, I breathed it, and then stood for several seconds overlooking the valley before me. I suddenly understood who the Shakers were and why they had chosen to live as they did. For the first time during my months of research, I felt that I could understand their love of order and beauty. It was as if I felt the heritage of that place pass through me (3.16.3).

A feeling of harmony with the world is one common element in these reports. 'At peace', 'warm towards everything', 'in harmony with everything', 'place pass through me' each suggest a self-non-self communion. Second, the experiencer feels more real. This vividness of presence is described as an inner tingling and quiet, as a spiritual moment – as a sense of reverence for time and place. Note, too, the importance of mood and physical environment. The person is quiet and receptive in the moment of contact; he or she seems unbothered and open. The environmental setting for these experiences all include natural features – water, vegetation, sunlight, birds. The size and beauty of the bridge and the extensive historical meanings of the museum indicate that immense overpowering environments or places grounded in significant history may provide important physical contexts for heightened encounter (Krawetz, 1975).

On the awareness continuum, heightened contact is the mode of encounter most tending towards person-world mergence (see Figure 14.2). The person feels joined and akin to the world. I overlap heightened encounter with noticing, since moments of the latter may lead to the former. Perhaps obliviousness and watching could also precede heightened contact, though probably not as often or readily.

Psychologists and social scientists in the past have often ignored or discounted heightened contact, arguing that it is 'subjective', 'mystical', 'illusory' or 'epiphenomenal' (Roszak, 1969). In the past two decades, however, a few researchers have begun to study heightened contact and accept it as a genuine and significant mode of human encounter (e.g. Searles, 1960, Tuan, 1961, Maslow, 1968 and 1969, Roszak, 1969 and 1973). The psychiatrist Searles (ibid.), for example, has termed heightened contact *relatedness with the environment.* For the psychologically healthy person, he argues, it can serve several important functions: assuaging various painful and anxiety-laden states of feeling; fostering a stronger realisation of self; deepening one's sense of reality; generating a stronger respect and acceptance of one's fellow men and women. The humanistic psychologist Abraham Maslow has studied what he calls *peak experiences*, which he believes all healthy persons have experienced sometimes in their lives; note the similarity to heightened contact:

> These experiences mostly [have] nothing to do with religion – at least in the ordinary supernaturalistic sense. They [come] from the great moments of love and sex, from the great esthetic moments (particularly of music), from the bursts of creativeness and the creative furor (the great inspiration), from great moments of insight and discovery, from women giving natural birth to babies or just from

loving them, *from moments of fusion with nature* (in a forest, on a seashore, mountains, etc.). . . (1968, p.10, italics added).

Figure 14.2: The Place of Heightened Contact on the Awareness Continuum

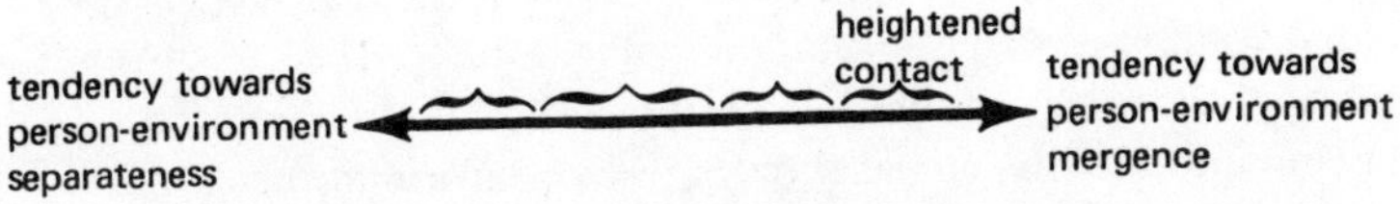

Heightened contact is relevant to behavioural geography because, as Maslow suggests, it can involve the physical environment beyond the person. The need is to understand more thoroughly how the physical environment can enhance or inhibit heightened contact and also ask if any sorts of educational programmes or techniques can work to help heightened contact happen.

Group observations indicate that heightened contact, like noticing, is unexpected and sudden. If this is the case, then perhaps education could foster a horizon of concern similar to that founding person-grounded noticing. This horizon would be a substantive area – for example, colour, plants, water, topography – in terms of which heightened contact could happen (Schwenk, 1961; Seamon, 1976a, 1978a; Grange, 1977). Chapter 16 considers this possibility further, particularly as it might foster a stronger ecological consciousness.[2]

Notes

1. For their appearance, coloured shadows require two conditions: (1) a coloured light source – e.g. a street lamp projecting a coloured light, and (2) a second light source, illuminating the shadow generated by the coloured light – e.g. a second street light emitting a white or yellow light. The shadow cast by the coloured light and illuminated by the second whiter light will be the colour's complement; thus, a blue light will produce an orange shadow; a green, red. For further details, see Goethe, 1970.

2. Heightened contact has bearing on the nature of encounter for many traditional and so-called 'primitive' societies. 'Here we find people', writes Lee (1959, p.164) 'who do not so much *seek* communion with environing nature as *find themselves* in communion with it. In many of these societies, not even mysticism is to be found, in our sense of the word. For us, mysticism presupposes a prior separation of man from nature; and communion is achieved through loss of self and subsequent merging with that which is beyond; but for many other cultures, there is no such distinct separation between self and other, which must be overcome. Here, man is *in* nature already, and we cannot

speak properly of man *and* nature.' This mode of encounter appears to be a kind of extended heightened contact. See Eliade, 1957; Nasr, 1968; and Moncrief, 1975.

Heightened contact may also have relation to the child's experience of nature. See Cobb, 1977.

15 BASIC CONTACT, ENCOUNTER AND AT-HOMENESS

> Last week I was walking from my dorm to the library, lost in thought, making plans for my parents' visit the coming weekend and where we might go to dinner. Just for a few seconds, I was able to watch myself walking up the hill, avoiding the puddles that had formed because of the rain we had last week. Something in me was watching for water puddles and guiding me around them. Even though I wasn't consciously aware of each puddle, something in my eyes and feet were working together. I'd see a puddle ahead and my feet would instantly hop over it or go around. All of this was happening as I was immersed in my thoughts. I didn't do a thing; it happened automatically – a group member (3.6.1).

Obliviousness, watching, noticing and heightened contact require *conscious* attention. The world, either inner or outer, takes on some degree of presence in the experiencer's awareness.

There is another kind of awareness: a preconscious attention, which like movement, arises from the body. I call this mode of encounter *basic contact*, and define it as *the preconscious perceptual facility of body-subject.* Basic contact works with the body, helping its actions to be in phase with the world at hand. Whether our more conscious awareness is experiencing a moment of obliviousness, watching, noticing or heightened encounter, basic contact is extending some amount of perceptual awareness. On the awareness continuum, basic contact is best defined as a wavelike structure running below the more conscious modes of encounter, always extending outwards some amount of prereflective attention (Figure 15.1).

Figure 15.1: The Place of Basic Contact on the Awareness Continuum

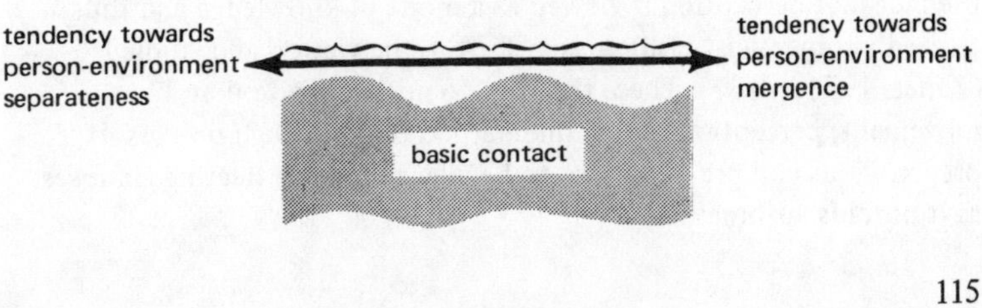

Examine the observation at the start of the chapter. The group member is deep in thought. He suddenly realises that there is a level of perception helping to guide him around puddles. Another group member, thinking about the day ahead, finds himself passing a car on the expressway (3.6.2). 'Some part of me', he says, 'was noticing and co-ordinating the road with the movement required to pass.' Basic contact is the 'some part of me' or 'something' which automatically guides behaviours. It, like movement, is precognitive: 'I was lost in thought'; 'I was oblivious to driving.'

Basic contact and movement are not separate experiential processes. Rather, they are part and parcel of a perception-movement reciprocity: basic contact assists movement which in turn brings about a new perceptual field. This reciprocity extends to person and world. No division exists between body and environment encountered: the two meet instantaneously in a person-world dialectic which sustains movements in the particular moment as it prepares movements for the next. As Merleau-Ponty explains:

> At each successive instant of a movement, the preceding instant is not lost sight of. It is, as it were, dovetailed into the present, and present perception generally speaking consists in drawing together, on the basis of one's present position, the succession of previous positions, which envelop each other. But the impending position is also covered by the present, and through it all those which will occur throughout the movement. Each instant of the movement embraces its whole span (1962, p.140).

The body, through its powers of movement and basic contact, integrates generalised patterns of habit with the uniqueness of the world at hand. We can understand Merleau-Ponty's claim (1962) that a philosophy of body-subject is already a philosophy of perception, since 'it is precisely through the body that we have access to the world: it is through this body that we have sensations whereby we experience the world' (Barral, 1965, p.119).

Limitations of both behaviourist and cognitive theories stem from the fact that perception is viewed as a mode of knowledge and thus treated in objectivist fashion as a 'stuff' to be viewed and studied (Zaner, 1971, p.131). These theories do not recognise that, like movement, perception can be understood only through *process.* It is preconscious and pre-objective, to be viewed only in fleeting glimpses as it unfolds automatically.

Perception, or *basic contact* as it is called here, is the mode of access through which the body meets the world and the generalised attitude of habit meets the particular environment at hand. Perception can not be spoken of experientially in terms of 'sense data', 'perceptual information', 'messages' and the like. It is a constant dialectic, an ever-present flow between person and world, allowing him to manage effectively simple and not so simple gestures, movements and tasks.

Basic contact is the essential foundation for more conscious modes of encounter. Because basic contact automatically synthesises our driving movements with the road ahead, we can turn our attention to the autumn foliage or notice skaters in the park as we pass. Or we can ignore the trip and think about the morning ahead or worry about a friend in the hospital. Basic contact, in harmony with the powers of habit, integrates the routine portions of our daily living. We can thus turn our attention to new and unfamiliar things. Alternately, we can choose to routinise our lives completely and grow inert and unfeeling in our encounter with the world.

Encounter and At-Homeness

Basic contact is an essential component of at-homeness. It provides a perceptual matter-of-factness, which comes to light only when the world is changed in some way. A familiar office feels strange because its blackboard, which was dirty, has been cleaned (3.14.2). A familiar street feels different because a tree along it has been cut (3.14.1). A change in the world as known brings itself to attention. That world seems different, strange, peculiar, and becomes a thing to be considered and figured out.

Rootedness houses basic contact. The perceptual field is automatically known in places of rootedness. The person can be oblivious to the world, yet his behaviours will be guided safely by the harmonisation between basic contact and body, person and world. Even in the most familiar environments, however, movements and perceptual field can go out of phase; the person is surprised or caught in an accident (3.7.2). One group member, pondering the future and oblivious, collided with a street sign:

> I didn't notice the 'no parking' sign on the edge of the sidewalk, and it caught my shoulder as I walked by. I jarred myself but was more surprised than hurt. It was an unexpected intrusion on my thoughts, a jolt from another world (3.7.1).

Beyond basic contact, at-homeness sustains a particular generalised attitude which permeates the person's daily existence and affects his modes of everyday encounter. On one hand, at-homeness fosters *habituality – the tendency of the person to take his everyday world for granted and notice little that is new or different.* Habituality is associated with the existential notion of *inauthenticity* – a stance of living in which the person does not deal with the world squarely, but experiences it as he has heard others say it is. 'Inauthenticity', says Relph (1976b, p.80), 'is an attitude which is closed to the world and man's possibilities. . . [it is] stereotyped, artificial, dishonest, planned by others, rather than being direct and reflecting a genuine belief system encompassing all aspects of existence.'

On the other hand, at-homeness may foster *openness – a situation in which the person strives for fuller understanding of the world because he feels comfortable and at ease.* Openness is related to the existential opposite of inauthenticity: *authenticity* – a mode of being in which the person accepts responsibility for his existence and seeks to be consistent and honest in his dealings with the world (Heidegger, 1962, p.68, Langan, 1959, pp.17, 21ff). Relph says:

> An authentic person is. . .one who is sincere in all he does while being involved unselfconsciously in an immediate and communal relationship with the meanings of the world, or while selfconsciously facing up to the realities of his existence and making genuine decisions about how he can or cannot change his situation (ibid., p.64).

Openness is a vehicle for authenticity. The person in openness looks with concern on his everyday world, its people, things and places. Openness involves a concerned attitude outwards; it is less associated with obliviousness (at least in its negative forms) and more related to watching, noticing, heightened encounter. The last two modes in particular relate to openness because they reveal unsuspected aspects of the world or foster intensified contact.

At-homeness fosters openness by conserving physical and psychic energies which can then be used for encounter and discovery. One group member, not feeling at home because of room-mate antagonisms, moved to a friendlier environment and felt happier. In the new situation, she had more energy to get involved in new experiences; to open herself to new possibilities:

> I definitely noticed that in the time I was living in this unpleasant situation there was no energy to spend on new activities. I was upset by the living situation and had no interest in doing anything other than the basic necessities. One of the reasons I had returned to college was to grow as a person – to try new things. But the apartment situation upset me. I had no wish to get involved in anything. I did just the minimal to keep me going. After I changed apartments and began living with people I liked, I felt more comfortable again. There was more energy. I could give muself to new things and take an active interest in life again. For example, I got involved in a pottery course and volunteered my help at a nursing home. I felt more free inside and could get involved with things outside myself (3.17.1).

At-homeness in other contexts fosters habituality. At-homeness guarantees familiarity and matter-of-factness. Life, if the person wishes, can proceed automatically and inauthentically, with a minimum of new encounters and contacts. Repetition and routine insulate response to the world. Daily living follows a comfortable monotony; the world is never questioned or looked at afresh. Habituality is associated with obliviousness and watching. Noticing is less probable, and heightened encounter – because of its unusualness and intensity – may be an impossibility.

Consider the following example of habituality. A group member, having lived several years in the same place, began to feel that her life was a mechanical process in which she took no active initiative. She felt a need to break away or live the same pattern forever:

> I had everything so easy. I had a job, friends, a nice place to live. The problem was that everything was *too* nice – life seemed stale. The same schedule day after day, the same people – everything was usually the same. One day I suddenly saw my situation and I was scared. I saw I could easily live like this the rest of my life. I thought, 'You've got to get yourself out of this rut.' I decided to make a change by returning to school (3.17.2).

Habituality and openness are both essential ingredients of a satisfying life. Habituality promotes order and continuity. We can not always explore the world for its fresh and unexpected aspects. Much of the time we must be practical and tend to the immediate needs at hand – getting to work, washing clothes, shovelling snow. We 'take hold' of the world quickly and effectively because of habituality. We conserve our energies and maintain our state of being. Time-space routines, body and place ballets are essential elements of habituality. Openness, in contrast, extends the person beyond himself. He contacts new aspects of the world and therefore grows as a person. Unknown aspects of the world become known. Realms of chaos and disorder are absorbed into an expanded sphere of at-homeness. The person extends his humanness.

Habituality is less useful when it stifles openness – when routine becomes so entrenched that the person forgets that life might be otherwise. Openness becomes potentially harmful when it extends the person beyond his reach – when it involves him in experiences and places which provoke danger or exertion beyond his capacities. Authenticity, says Heidegger, is an aim of living, not an end in itself: 'authentic existence, in fact, can only be something of an ideal, a direction to aim at amidst the dark reality of the dissimulation of everyday life' (Langan, 1959, p.25). At-homeness, in its dual powers to foster both habituality and openness, is the backdrop out of which authenticity is possible. In this sense, an understanding of at-homeness and encounter clarifies the nature of an authentic mode of being and helps the person who wishes to move towards it.

16 IMPLICATIONS FOR ENVIRONMENTAL THEORY AND EDUCATION

> [A] 'knowledge problem'. . .has haunted western philosophy. . . insistently since Descartes. The relations between knower and known are problematic with us because we have grown so peculiarly stupid about the way experience really happens. Even (or perhaps especially) in the work of our leading modern philosophers, discussion of the sense life is incredibly insipid; experience has neither power nor complexity for them. . .our philosophy often trails off into much bookish discussion about something called 'sense data', conceived of abstractly as a uniform species of evidence that politely registers its arrival and then waits about to be accounted for in clever epistemological schemes – Theodore Roszak (1973, p.83).

Standard psychology and philosophy generally reduce encounter to perception, which is then discussed in causal, mechanistic terms (Keen, 1972, p.90). Behaviourists, on the one hand, interpret perception as a chain of responses to the external environment; the significant perceptual structure – the environment as stimulus – exists *out there* in the world. Cognitive theorists, on the other hand, see perception as an information-chain that works in cybernetic fashion; the significant perceptual structure – a cognitive deciphering apparatus in the mind – lies *inside* the person.

Phenomenology looks away from these arbitrary interpretations and back towards perception as an experience. Phenomenologically, the essential perceptual structure can not exist inside the head nor out in the world. Perception is a dynamic inner-outer relationship. It is the variegated and fluctuating bond of attention between person and world, body and environment, sometimes stronger in its union, sometimes weaker.

Perception is therefore an impoverished term to describe the variegated ways in which people attend to their world. It (or *basic contact*) more appropriately describes the preconscious awareness of body-subject. This is the meaning Merleau-Ponty (1962) had in mind when he incorporated the word into the title of his major work, *Phenomenology of Perception. Encounter* is a better description for the ways we attentively meet the world. Encounter is a multifaceted ebb and flow of attention and involves all shades of obliviousness, watching, noticing and heightened contact. Beneath these more conscious attentive modes is the steady stream of basic contact, which in all but the most oblivious of moments keeps body and world, movements and surroundings, in smooth attunement. The sum is depicted in Figure 16.1.[1]

Figure 16.1: The Awareness Continuum as a Whole

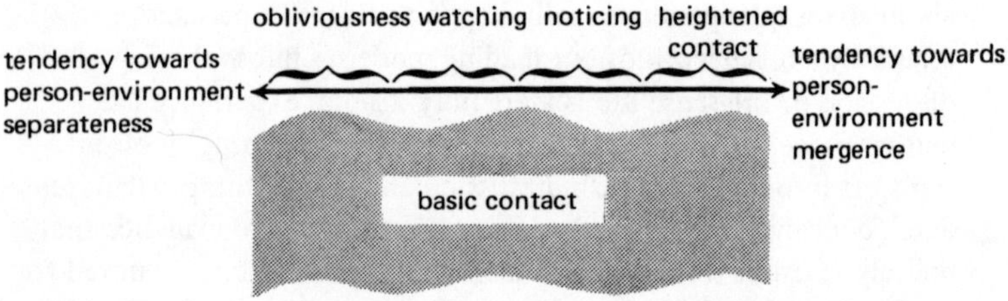

Conventional Perception Research: The Case of Landscape Assessment

Work in landscape assessment demonstrates well how the variety of encounter modes has been ignored in much conventional perceptual research. This work attempts to determine the precise qualities of the physical environment which particular individuals and groups find attractive.[2] In what kinds of houses and neighbourhoods do people ideally like to live (e.g. Michelson, 1966)? How does the presence and arrangement of water, vegetation or man-made features affect a person's appraisal of a particular natural landscape (e.g. Brush and Shafer, 1975)? Do people with different personalities and different socio-economic backgrounds perceive the same landscape differently (e.g. Craik, 1975)?

Operationally, this research usually presents subjects with actual landscapes, or shows them simulated environments by means of photographs, slides or scale models. Subjects are asked to describe and evaluate these real or represented landscapes by means of adjective checklists, semantic differentials, or some similar measuring device. Statistical tests are applied to the resulting data to determine categories of preference and variations in preference among individuals and groups. This research has had useful consequences for environmental planning and policy decisions because it provides information on landscape preferences, which can be used in decisions on land use and environmental design (Zube, 1973).

Phenomenologically, research in landscape assessment is open to considerable criticism because it loses sight of the many ways in which landscape can be encountered by the experiencer. Encounter is *contextual* – in different moments we experience environment differently. When I arrive at Mt Monadnock, tired after my lengthy car drive, I ignore the environment for which I have sacrificed an entire morning. I begin hiking, however, and shortly feel refreshed. The environment opens itself to me, and I begin to notice things to which I was oblivious a short time before.

Studies in landscape preference reduce the multifaceted modes of encounter to the artificial situation of person actively evaluating a real or simulated landscape. The subject plays the fabricated role of a person-grounded noticer; he brings attention to the landscape only because a researcher requests it. The subject actively peruses the environment – studying it and judging it, deciding what he likes and does not. His judgements are affected by the descriptive characteristics to which the questionnaire or semantic differential draws his attention. Many of these qualities he might never notice if he were there with the environment in the natural attitude.

Studies of landscape preference transform encounter into an exercise of evaluation. Information provided by the subject may have little correspondence to the significance of that landscape for the same subject when he is there on a particular day actually looking at it, recreating in it, passing through it, or being oblivious to it. These studies fail to grasp the fluidity of encounter and transcribe its variety into static, objectivistic terms. In general, this research suggests that a physically attractive environment will be an encountered environment. Brush and Shafer (ibid.) conclude, for example, that the greater the amount of vegetation and water, and the less the amount of non-vegetation, then the higher the chances that a particular landscape be preferred. Physical characteristics of the environment have some bearing on noticing and heightened contact, but they are not necessary or sufficient in themselves. As we have seen, other characteristics such as inner state and past experience may have as strong or stronger bearing on encounter.

In suggesting that scenery is a resource and so needs inventory and measurement, Zube argues that 'we must identify those aspects of scenery which mean something to the broadest range of people' (Zube, 1973, p.130). The awkwardness here is that meaning does not necessarily foster encounter: just because I rank a particular landscape 'beautiful', 'challenging', 'clean' and 'green', does not mean that on the particular day I hike through that landscape it will necessarily have more significance for me than another landscape which I ranked 'angry', 'bare', 'forbidding' and 'stony'.[3] Ugliness and contrast may as often foster encounter as the beauty and attractiveness which most landscape assessment studies highlight and work to institutionalise. No doubt it is important to protect landscapes that are 'beautiful', 'clean', 'grassy', 'green', 'hilly', 'natural', 'peaceful', 'pleasant', 'sunny' and 'tree-studded'.[4] These kinds of environments are clearly important to many people and the landscape assessment studies serve a valuable function in that they provide empirical validation for that importance. 'Quantitative data', writes Zube (1973, p.130), 'carry more weight with environmental decision-makers than do arguments based on "emotion" or personal feelings.'

On the other hand, the preservation of landscapes which possess the attributes people have labelled important does not mean that those people as experiencers will always encounter these environments in the way they have labelled them. In this sense, we must recognise the complementary side to landscape meaning: the medium of intercourse between landscape and experiencer which may lead to noticing, watching, heightened contact – or obliviousness.

Besides preserving attractive, stereotyped landscapes, then, we should also be concerned with programmes which would sensitise people to *all* environments – be they superficially appealing or not. Such education would, first, introduce people to the variety of ways in which they encounter the physical environment, and, second, create techniques that would enhance their abilities to see and experience. Perceptive persons in all ages have frequently discovered beauty and depth in things which at first sight appear mundane, inconsequential or unpleasant (Roszak, 1973). Can we foster such sensibilities in ourselves and discover meaning and beauty in the seemingly commonplace? Such education would be particularly valuable in cities, where so-called attractive environments are less common.

Education and Environmental Encounter: Delicate Empiricism

Education in environmental encounter is concerned with those modes of awareness which extend the person's understanding of the world around him. Watching in new ways, noticing, heightened contact, openness and authenticity are therefore important. The person works to discover the world for himself,; to meet it authentically: his aim is to see the world as it is in its own fashion – not as other people tell him it is. This mode of encounter is reflected in Relph's description of authentic experience of place. The place is met as a thing in itself before interpretation and colouring by preconceived thoughts and cultural filters:

> An authentic attitude to place is. . .understood to be a direct and genuine experience of the entire complex of the identity of places – not mediated and distorted through a series of quite arbitrary social and intellectual fashions about how that experience should be, nor following stereotyped conventions (1976b, p.64).

Many approaches and techniques for fostering openness and authentic encounter exist. Painting, dance and other artistic means are perhaps one of the most valuable vehicles because they promote emotional contact with aspects of the world.[5]

One method appropriate for education emphasising cognitive understanding is the scientific work of Goethe, whose use of interpersonal verification was discussed in Chapter 2. Goethe was profoundly interested in the natural world and conducted explorations of such phenomena as light, colour, plants, rocks and weather. Discouraged by the theorising, analysing and measuring of conventional science of his day, he developed a mode of investigation that he called *higher contemplation (Höhere Anschauung*) or *delicate empiricism (zarte Empirie*). Goethe's method sought to explore things experientially – to foster a moment of intensified encounter through which the thing could be discovered and understood as it was *in itself* before any observer had defined, categorised or labelled it.

The prime aim of delicate empiricism, therefore, is to understand the thing through experiential contact. 'Pure experiences', Goethe wrote, 'should lie at the root of all physical sciences.'[6] These 'pure experiences', Goethe argued, are to be found in a moment of sudden insight in which the student sees the thing in a deeper and more vivid way. 'One instance is often worth a thousand', wrote Goethe, 'bearing all within itself.' The student's task is to encounter the thing intimately – to penetrate its aspects through the powers of human seeing and understanding.

Delicate empiricism uses watching and noticing focused on a particular thing. The ultimate aim is a moment of heightened contact in which person and thing emerge. Care is involved: the student is asked to extend a sense of reverence towards the thing. He should not look at it dispassionately or callously as a physiologist might study a rat. Nor can he manipulate and master the thing or make it into what he thinks it is. Rather, Goethe believed that 'natural objects should be sought and investigated as they are and not to suit observers, but respectfully as if they were divine beings.'

Delicate empiricism works to channel a spirit of openness and concern outwards towards a particular thing through training and heightening encounter. Delicate empiricism is similar to phenomenology in that the student must work to establish an atmosphere of receptivity to the thing. It is also closely related to Heidegger's spirit of dwelling in that it helps the person see a thing more deeply and clearly and thereby feel more at home with and responsible for the thing.

Foundational Ecology

One subfield of behavioural geography is concerned with human attitudes towards nature and the physical environment.[7] Delicate empiricism and similar approaches have bearing on this subfield because they foster what Grange (1977) has called a *foundational ecology* – an attitude towards the physical environment grounded in reverence and concern. Foundational ecology, says Grange, is considerably different from the predominant ecological consciousness of today, which he calls *dividend ecology* – an environmental attitude founded in fear and sense of economic threat. Dividend ecology regards

> the interaction of humankind and nature solely from the perspective of investments and returns. The slogans of dividend ecology are familiar: 'Don't litter', 'keep your campsite clean', 'Pitch In', and so on and so forth. Dividend ecology has a simple message: if we continue to destroy our environment, we will perish. Its motive force is fear, being largely a negative movement that seeks to restrain our greed and diminish the aggression with which we attack nature. This way of understanding ecology can do little in the long run, for it only serves to reinforce the basic mode of consciousness that brought on our environmental disaster (ibid., p.136).

Instead, says Grange, we must foster a foundational ecology – a sense of environmental obligation arising out of kindliness and respect for the natural world. Making reference to Heidegger, Grange argues that foundational ecology is one aspect of dwelling; it is

> the effort to structure our modes of dwelling so that they reflect an essential and authentic way of being human. That way is an existence

> that opens itself to nature rather than aggressively reconstructing it according to personal ends. And the purpose of this way of dwelling is not to preserve 'wilderness' for our children or any of the hundred other reasons given by dividend ecology. Rather we seek to dwell so that we can move nearer to that which resides hidden at the center of ourselves (ibid., p.148).

Foundational ecology, says Grange (ibid.), does not mean that we renounce technology or return to a primitive condition. Rather, this attitude might change our way 'of building and constituting the world' and allow us to use technology in a way that is more humanly and environmentally constructive (ibid., p.147). We return therefore to Heidegger's belief that dwelling must precede building, which also means that at-homeness must replace homelessness. 'To come home', concludes Grange (ibid., p.148), 'is to undertake a way of relating to nature that allows nature to show itself to us and that encourages us to abide and take up residence in that meaning.'

Foundational ecology is fostered practically by approaches like Goethe's delicate empiricism, which sensitise the student to one particular aspect of the natural or human worlds – clouds, plants, rocks, places or whatever. One significant example here is Theodore Schwenk's *Sensitive Chaos* (1961), an application of Goethe's approach to water. In the past, Schwenk believes, people treated water with reverence, which led to a right but unselfconscious ecological relationship with it. Over time there occurred a change in attitude and today man looks 'no longer at the *being* of water but merely at its physical value' (ibid., p.10). Schwenk works to demonstrate through words, drawings and photographs that water has an essential character, described by such features as the wave, vortex and vortex ring. He suggests that by looking at and studying water in this way people might again feel concern and care for their water resources.

Notes

1. Clearly, the modes of encounter discussed here are not the only modes. What, for example, are 'looking', 'looking at', 'observing' or 'studying'. These modes as well as others are also aspects of human awareness and future phenomenologies of encounter could usefully seek distinctions and clarification.

2. For one overview of this work, see Zube *et al.* (eds.), 1975.

3. These words are taken from the landscape adjective checklist used by Craik (1975, pp.138-9).

4. Craik found that of the 240 adjectives comprising his list, these words were

most frequently checked by his subjects.

5. One example is Kimon Nicholaides' *The Natural Way to Draw: A Working Plan for Art Study* (1949), which provides a series of drawing exercises that improve the student's artistic abilities as they intensify contact with the thing drawn. These exercises, applied to environmental themes, could do much to kindle love of nature.

6. The source of this and other quotations from Goethe can be found in Seamon (1978a).

7. For an overview of this subfield, see Ittelson *et al.* (1974, pp.17-59).

Part Five

SEARCHING OUT A WHOLE

In a life span, a man now –
as in the past – can establish
profound roots only in a
small corner of the world
– Yi-Fu Tuan (1974, p.100).

17 MOVEMENT AND REST

> I suggest we think about places in the context of two reciprocal movements which can be observed among most living forms: like breathing in and out, most life forms need a *home* and *horizons of reach* outward from that home. The lived reciprocity of rest and movement, of territory and range, of security and adventure, of housekeeping and husbandry, of community building and social organization – these experiences may be universal among the inhabitants of Planet Earth. Whether one thinks on the level of ideas themselves, or of social networks, or of 'home grounds', there may be a manner in which one can measure and study the reciprocity of home and reach in all of them – Anne Buttimer (1978, p.19).

Phenomenology explores parts to understand wholes. The whole here is everyday environmental experience. It has been explored in terms of movement, rest and encounter, which can be represented as a whole by the triadic structure of Figure 17.1. A *triad* is 'a union or group of three' (*Webster's Seventh*, 1963, p.945). I call this representation a triad (rather than a triangle) because the term suggests a working relationship among the parts – as in a chord triad of music.[1] Borrowing from past chapters, the following can be said about this *triad of environmental experience:*

- that people become bodily and emotionally attached to their geographical world;
- that this nexus of attachment is at-homeness;
- that at-homeness sustains a taken-for-granted pattern of continuity, expectedness and order;
- that as people move and rest in their geographical world they also encounter it;
- that encounter varies in the degree to which people are part of or apart from their world.

The last chapters of this book consider in further detail the various links among movement, rest and encounter. The aim is to gain a better understanding of their threefold relationship – to point to ways in which themes in earlier chapters interweave in wider patterns of

Figure 17.1: Triad of Environmental Experience

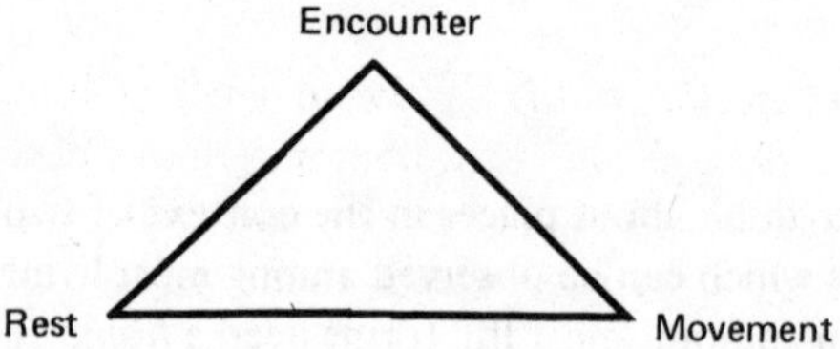

meaning. In this way, one better understands everyday environmental experience as a whole.

A phenomenology of any experience or phenomenon can never be judged complete. The student, as he proceeds, discovers wider and finer meanings and structures which provide potential new paths for exploration. These last chapters discuss several themes which did not arise directly from the environmental experience groups. Rather, these themes have grown out of my own reflections and discoveries which arose in the process of writing preceding chapters. In this sense, the following patterns and themes have not all been explored intersubjectively and are therefore tentative. Perhaps in time they may become the focus of further group exploration.

The Dialectic between Movement and Rest

Movement and rest are not isolated phenomena: they exist together in dialectic.[2] There is a continuous tension between the two which leads to a series of resolutions. 'Man', says the architect Aldo van Eyck (cited in Norberg-Schulz, 1971, p.33), 'is both center bound and horizon bound.' Movement leads to rest which in turn leads to movement. This dialectic can be represented as in Figure 17.2.

Rest is associated with centre, home and at-homeness. It points to a basic human need for spatial and environmental familiarity and order. Rest anchors the person's present in his or her past; it maintains

Figure 17.2: The Dialectic between Movement and Rest

experiential continuity. Security, privacy, quiet, passivity, contemplation and other similar qualities often have their context in rest.

The deepest experience of rest is dwelling, which brings people together and people and nature together in terms of place. The world of dwelling involves regularity, repetition and cyclicity all grounded in care and concern; it is, as Jager says, a 'round world':

> The round world of dwelling offers a cyclical time, that is, the recurring times of seasons, of the cycles of birth and death, of planting and harvesting, of meeting and meeting again, of doing and doing over again. It offers a succession of crops, of duties, generations, forever appearing and reappearing. It offers a place where fragile objects and creatures can be tended and cared for through constant, gentle reoccurring contacts (1975, p.251).

Movement, in contrast, has links with horizon, reach and unfamiliarity. It is associated with such active qualities as search, newness, exploration, alertness and exertion. A person, through movement, extends his knowledge of distance, place and experience; he becomes familiar with spatial and experiential horizons that were undisclosed or obscure before. Movement helps the person to assimilate places and situations into his world of familiarity. In this sense, movement widens the sphere of at-homeness and dwelling.

Jager (ibid.) discusses movement in terms of *journey*, which he finds a common theme not only in travelling, exploring and sight-seeing, but also in intellectual, artistic and spiritual efforts. The journey carries the person away from his stable world of dwelling. It gives him a sense of forward and back, past and future; and moves him outward along a path towards confrontation – with places, experiences, ideas:

> Journeying forces [the] round generative world of [dwelling] into the narrow world of the path. The path offers the progressive time of unique and unrepeatable events, of singular occurrences, of strange peoples and places to be seen once and possibly never again. . . Journeying breaks open the circle of the sun and the seasons and forms it into a linear pattern of succession in which the temporal world shrinks to a before and after, to backward and forward. Here the beginning is no longer felt to lie in the middle but instead appears placed behind one's back. The future makes its appearance straight ahead, making possible *confrontation* (ibid.,

p.251, italics in original).

Movement and rest, because of their dialectical nature, each incorporate aspects of the other. They are not mutually exclusive but often encompass qualities more often associated with the opposite. The afternoon constitutional, for example, first suggests movement. As an experience, however, this daily walk through a regular sequence of paths and places typically involves no adventure or unfamiliarity. It may, however, invigorate the walker and renew his depleted energies. Here, an action which in appearance suggests movement provides an inner function associated with rest. Rest and movement exist together, and each shares aspects of the other.

The dialectic between movement and rest extends over a variety of spatio-temporal spheres. In one sense, our geographical existence can be likened to a continual series of stops and starts at all scales of time and space. I sit quietly in my chair for an interval and then *reach* for a glass of sherry at my side; the housewife works for the morning in the kitchen and *moves* to the porch to crochet; the family on their weekly Sunday drive travel for an hour and then *stop* for lunch; the storekeeper lives in the apartment above his shop for fifty weeks of the year but each May *takes* a two-week vacation to his native Greece. Life, described in this way, is a series of pendulum swings between movement and rest. Through movement, people leave the taken-for-grantedness of place or situation and extend their horizons elsewhere; through rest they return to particular centres and collect themselves in preparation for future ventures outward again. Each requires its opposite in order for itself to be so.

Both Bachelard (1958) and Relph (1976b) speak of the relation between movement and rest as an inside-outside dialectic. Bachelard (cited in Relph, ibid., p.49) explains that 'outside and inside form a dialectic of division, the obvious geometry of which binds us. . . Outside and inside are both intimate – they are always ready to exchange their hostilities.' Relph (ibid., p.49) gives the simple example that 'we go out of the city into the countryside, yet return again into the city.' A place of rest, the rural landscape, in time loses its attractiveness and the person returns home to another place of rest. What is experientially 'inside' for a time becomes 'outside'. This frequent 'exchange of hostilities', as Bachelard calls it, is a continual and inescapable aspect of people's geographical existence. In part because of this exchange, people gain both stability and stimulation in their day-to-day lives.

Particular places have their own particular threshold of inside and outside, staying and leaving. I arrive at my office at eight and leave at five. If I stay beyond that hour, I normally feel tired, anxious or uncomfortable. Around five o'clock, in other words, my office loses its quality of insideness, which is transferred to another place – my home. Similarly, I take a coffee break at the nearby luncheonette but can usually stay there no longer than a half-hour. Beyond that time, I begin to feel out of place and automatically return to my office, which regains the quality of insideness held by the luncheonette the half-hour before.

The whole of a person's life can be viewed through the dialectic of movement and rest, inside and outside, dwelling and journey. Changes in place – from hour to hour, day to day, year to year, early adulthood to middle age – can all be interpreted in terms of a need to move and rest, to stay in a particular place for a time and then move elsewhere. These temporal thresholds arise partially out of feeling-subject, which after a certain time begins to feel discomfort, boredom, wanderlust, or some similar emotional push or pull which moves the person to another place or situation. Body-subject, at least in shorter periods of movement and rest, also has some role, automatically moving the person elsewhere because of a particular time-space routine.

The body, therefore, has its own sense of time, habitually moving when a particular temporal threshold is reached and emotionally provoked by feeling-subject when that threshold is overextended. Occasionally, cognition may interrupt this threshold process, ordering the person to stay in place longer than he normally would, or deciding to leave that place sooner than usual. More frequently, however, movement and rest, inside and outside change places automatically, and the person's daily routines proceed with a minimum of conscious directedness.

Imbalance and Balance between Movement and Rest

An imbalance of movement or rest in a person or group's life provokes awkwardness, discomfort or stress. As Relph (ibid., p.42) suggests:

> Our experience of place, and especially of home, is a dialectical one – balancing a need to stay with a desire to escape. When one of these needs is too readily satisfied we suffer either from nostalgia and a sense of being uprooted, or from the melancholia that accompanies a feeling of oppression and imprisonment in a place.

Besides the oppression and imprisonment to which Relph refers, an excess of rest may also be associated with isolation, withdrawal, drudgery and provincialism. Robert Coles' portrait (1967) of Appalachian mountaineers is one example of a people extremely bound in time and space. Their existence, in one sense, is too sheltered and they find it difficult to cope with events and people that intrude from the outside. On the other hand, an excess of movement may preclude adequate rest and be associated with the uprootedness and nostalgia that Relph mentions, as well as injury, exhaustion, home-sickness, or overextension. Coles' description (1967) of Southern migrant workers portrays a situation of movement at the expense of rest. Because they have no roots, these people know only minimal order in terms of place and home. Coles (ibid., p.116) concludes that this uprootedness fosters a life of wretchedness – especially for the children:

> Even many animals define themselves by where they live, by the territory they possess or covet or choose to forsake in order to find new land, a new sense of control and self-sufficiency, a new dominion. It is utterly part of our nature to want roots, to need roots, to struggle for roots, for a sense of belonging, for some place that is recognized as *mine*, as *yours*, as *ours*. . . It is quite another thing, a lower order of human degradation, that we also have thousands of boys and girls who live entirely uprooted lives, who wander the American earth, who even as children enable us to eat by harvesting our crops but who never, never can think of any place as home, of themselves as anything but homeless (italics in original).

Movement and rest, inside and outside, dwelling and journey, therefore, each suggest and sustain the other. A satisfactory life can be said to involve some balance between these opposites. Perhaps Jager (ibid., p.249) has summarised this balance best, portraying it as a reciprocity and interpenetration between dwelling and journey:

> Journeying grows out of dwelling as dwelling is founded in journeying. The road and the hearth, journey and dwelling mutually imply each other. Neither can maintain its structural integrity without the other. The journey cut off from the sphere of dwelling becomes aimless wandering, it deteriorates into mere distraction or even chaos. . . The journey requires a place of origin as the very background against which the figures of a new world can emerge. The hometown, the fatherland, the neighborhood,

> the parental home form together an organ of vision. To be without origin, to be homeless is to be blind. On the other hand, the sphere of dwelling cannot maintain its vitality and viability without the renewal made possible by the path. A community without *outlook* atrophies, becomes decadent and incestuous. Incest is primarily this refusal of the path; it therefore is refusal of the future and a suicidal attempt to live entirely in the past. The sphere of dwelling, insofar as it is not moribund is interpenetrated with journeying (italics in original).

For different individuals, groups and historical times, the exact nature of the balance between movement and rest, dwelling and journey no doubt varies, but its presence in some form always exists. The student can perhaps come to understand this balance best by first exploring its nature in his own life situation. He can then consider balances in other people's lives and ask how they compare and contrast with his own.

Notes

1. The notion of triad (as well as *tetrad*, which will be discussed shortly) owes its origin in part to the work of the English philosopher Bennett (1966), who has developed a method which he calls *systematics* – a way of exploring wholes through the qualitative significance of number. On triads and tetrads, see ibid., pp.23-37.

2. Dialectic has a considerable range of meanings; here, I use the term to mean 'the dialectical tension between two interacting forces or elements' (*Webster's Third*, 1966, p.623). Bennett (ibid., pp.18-23) terms this dialectical relationship a *dyad*, which he says 'is in a state of tension' (p.23).

18 THE TRIAD OF HABITUALITY

People are their place and a place is its people, and however readily they may be separated in conceptual terms, in experience they are not easily differentiated – Edward Relph (1976b, p.34).

People encounter the world as they move and rest, dwell and journey. Encounter can be joined with movement and rest by inverting the awareness continuum of Figure 16.1 and creating the tetradic structure of Figure 18.1 below. A *tetrad* is 'a group or arrangement of four' (*Webster's Seventh*, 1963, p.913). Like the triad, this symbol should be seen as representing a dynamic process rather than a static form (Bennett, 1966, pp.19-37).

Figure 18.1: The Tetrad of Environmental Experience

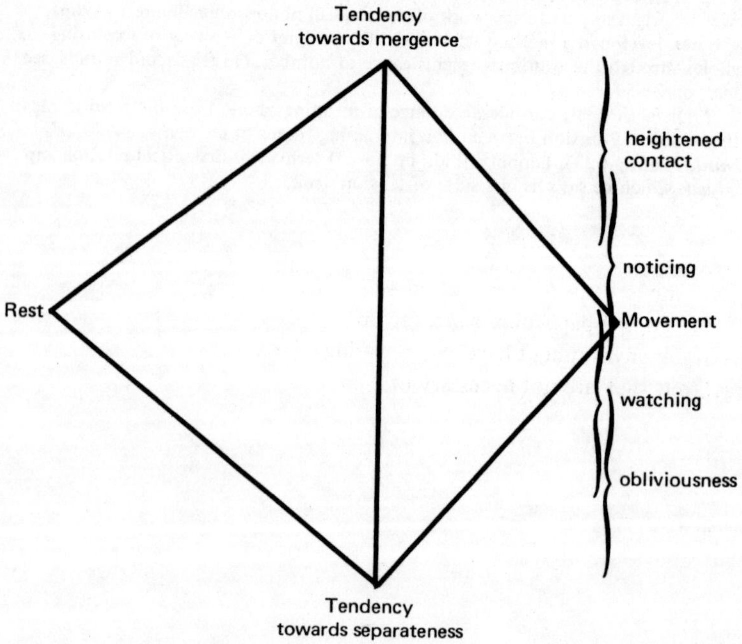

If one examines this *tetrad of environmental experience*, he notes that it is composed of two separate triads marked out by movement, rest and the tendency towards separateness on the one hand; and movement, rest and the tendency towards mergence on the other (Figure 18.2). The upper triad can be thought of as a *triad of openness* because it more often incorporates modes of encounter – heightened contact, noticing grounded in the person – through which one reaches out to the world at hand and discovers more about it. It will become clear that experiential and environmental education have bearing on this triad (Chapter 20).

In contrast, the lower triad can be called the *triad of habituality* because it includes modes of encounter – obliviousness, undisciplined watching, noticing grounded in the world – that are more often associated with taken-for-grantedness and routine. This triad has implications for environmental design, especially as it has relation to place ballet (Chapter 19).

Figure 18.2: The Triads of Openness and Habituality

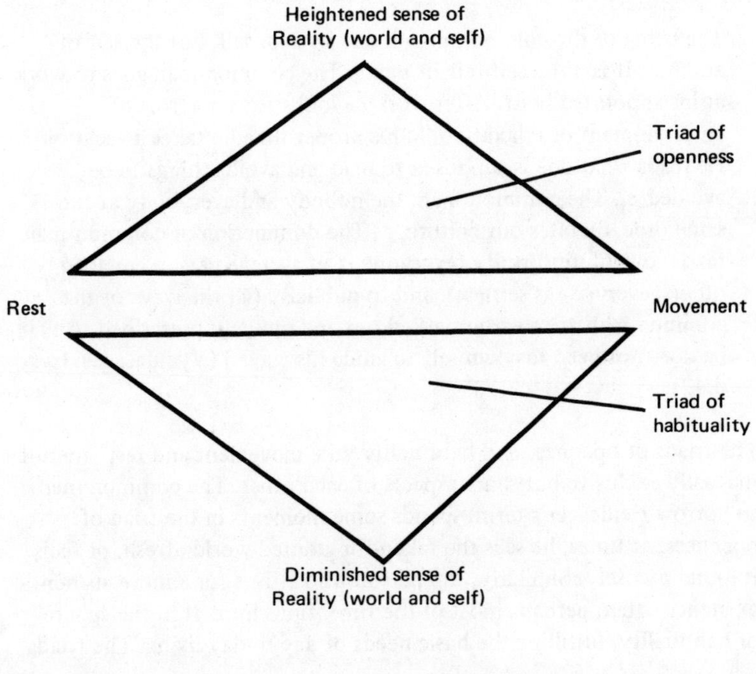

Two Triads and the Common Man

In Heideggerian terms, the openness triad reflects a more authentic existence than the triad of habituality. Through openness, the person widens his grasp of the world, but in a different way than through movement (which extends the person by introducing him to new places and situations). Rather, openness widens the individual's horizons in that it fosters a more refined understanding and greater concern for the world; it works 'as a removal of concealments and obscurities, as a smashing of barriers with which [being] bars itself from itself' (Heidegger, cited in Vycinas, 1961, p.42).

In contrast, the habituality triad is the world of habitual movement and rest; it is the realm of body-subject, feeling-subject and routine. Here, the person need pay only minimal attention to the world at hand. The attention he or she does give the world is usually expressed through watching or noticing fostered by the environment. Life proceeds with a minimal degree of change or newness. For Heidegger, this triad is associated with the world of *common man*, who does not encounter the world for himself but gains his understanding though the expressions and dictates of others:

> The being of the common man is not its own self, but the self of another. It is not a self-reliant being. The common man goes to work at his appointed hour. After work he looks for some sort of entertainment or relaxation; in his proper time he takes a vacation. He reads what one is supposed to read and avoids things to be avoided. . . The common man, the nobody and everybody at the same time, dictates our culture. . . The domination of common man tends toward uniformity (everyone is an average person equal to others; everyone is selfless), and to publicity (all the ways of the common man are clearly marked out and publicly prescribed so that he does not need his own self to guide his ways) (Vycinas, 1961, p.42).

The triads of openness and habituality, like movement and rest, are not mutually exclusive but share aspects of each other. The common man, to borrow Heidegger's term, spends some moments in the triad of openness: at times, he sees the taken-for-granted world afresh, or feels it more intensely. Similarly, the person who strives for a more authentic existence often, perhaps most of the time, finds himself in the sphere of habituality, fulfilling the basic needs of day-to-day living. The triads of openness and habituality reflect a dialectic between two different modes of awareness and encounter. Each requires the other, and an authentic existence can proceed only because the person recognises his frequent inauthentic existence. As Vycinas (ibid., p.42) explains, 'the inauthentic way of existence is needed to provide the grounds in which the authentic mode of being can be built. Authenticity is nothing more than modified inauthenticity.'

The Triad of Habituality and Placelessness

The triads of openness and habituality are significant to behavioural geography because they reveal two complementary ways to study environmental behaviour and experience. The habituality triad points to a way of living that is very much taken for granted and mechanical. The researcher can best study this domain if he focuses on the prereflective, matter-of-fact aspects of daily living. The common man typically resents change or interference in the well established patterns of his life. The nature of the world – particularly the physical environment – becomes important, for it, rather than its people, is often more easily modified and changed for the better. Through this change, places might become more authentic.

In the past, explains Relph (1976b, p.68), the common man frequently created authentic places and he typically did this *unselfconsciously.* His ability to manipulate the environment was generally limited technologically; he therefore automatically created places of human scale built by hand and animal power from local materials. Illustrations, says Relph (ibid., p.68), are found in the architecture of primitive and traditional peoples, for example, the preindustrial villages of Europe. The end result is

> places which fit their context and are in accord with the intentions of those who created them, yet have a distinct and profound identity that results from the total involvement of a unique group of place-makers with a particular setting.

In any era, says Relph (ibid., p.71), there are always some places that are 'lived-in and used and experienced' and therefore authentic. Today, however, Relph argues, because of a powerful technology, mass communications and governmental centralisation, the places of common man are more often inauthentic. They are the placelessness seen in the frequent disorder and monotony of suburban development; the alienating public spaces that belong to everyone and no one at the same time; the look-alike Holiday Inns or some similar corporate chain that has little concern for locality or its people, other than to drain away money and resources.

Places today are less authentic because they are less often built by individual people who have a stake in place and therefore feel kindness and concern for it and the wider locality. Attachment for place arises out of being in and living in a place. Nameless corporations or public agencies are not individual people and can muster no such attachment; their concern for place is reduced to economic motivation or the anonymous 'public welfare'. As a result, says Relph (ibid., p.140), 'there is little scope for the development of more than a casual sense of place because the identities of places are merely the product of fabrications or of local associations of universal and placeless processes.'

Can the growing placelessness of the world be tempered? Are there practical implementations that would restore the individual's and group's sense of place? One notion of value is the place ballet whose various aspects have been discussed in past chapters and now can be integrated into a whole.

19 PLACE BALLET AS A WHOLE

> Under the seeming disorder of the old city, wherever the old city is working successfully, is a marvelous order for maintaining the freedom of the city. It is a complex order. Its essence is the intricacy of sidewalk use, bringing with it a constant succession of eyes. This order is all composed of movement and change, and although it is life, not art, we may fancifully call it the art form of the city and liken it to a dance – not to a simple-minded precision dance with everyone kicking up at the same time, twirling in unison and bowing off en masse, but to an intricate ballet in which the individual dancers and ensembles all have distinctive parts which miraculously reinforce each other and compose an orderly whole – Jane Jacobs (1961, p.50).

To regenerate unique and authentic places today requires, first, a self-conscious understanding of human environmental experience; and, second, a means of translating this understanding into action, through life-style and building design. Realisation that dwelling precedes building, that community almost certainly arises out of place might lead to a landscape reflecting diversity and a harmonisation of people, environment and values.

Such change cannot happen suddenly; it may never happen. For those people who see placelessness as a neutral event or as a sign of equality, prosperity and progress, such change may seem retrogressive or an irrelevant nostalgic ideal. Yet if bodily and emotional requirements, as well as the need of dwelling, point to place as more conducive experientially than placelessness, it can well be asked how more places can be generated practically.

In this book, the most valuable practical notion is place ballet, which, I have suggested, is a foundation of at-homeness and dwelling. The groundstone of place ballet is the coming together of people's time-space routines and body ballets in terms of space. Additional, less regular participants may be drawn to place ballet, but its crux is the prereflective bodily regularity of routine users. Place ballets can involve a room, a corridor, a city park, a block of houses, a shopping mall – even an entire town, city or region. Place ballets can be a bastion of activity in an empty and dull larger landscape, or they may interpenetrate in interlocking, wider wholes to create an environment of vitality, motion and sense of community.[1]

This chapter works to suggest the essential character of place ballet. I speak of six qualities: *attraction, diversity, comfortableness, invitation, distinctiveness* and *attachment.* The argument is that any place ballet involves to some degree all or some of these qualities.

Discussion in this chapter is based primarily on my own reflections and is therefore tentative.[2] Notions and ideas eventually need clarification and correction by phenomenological studies of specific concrete place ballets occurring at different environmental scales in different times and places. The reader makes best use of this chapter if he or she considers themes in relation to specific place ballets with which he or she is familiar. Do they reflect the qualities suggested here and if so, to what degree? Do they involve additional qualities? Do particular features of the physical environment work to sustain or hinder the place ballet?

Attraction

For a place to have people, it must attract them to itself. The luncheonette offers coffee and food; the shop sells merchandise; the apartment house brings home its dwellers; the busy street provides movement and events to which the would-be watcher is drawn. No one normally goes somewhere that has no kind of offering. 'Most of us', says Jacobs (1961, p.129), 'identify with a place. . .because we use it, and get to know it reasonably intimately. We take our two feet and move around in it and come to count on it.' Using a place is a key to all place ballets.

A place which attracts is a focus of some activity. 'Counting on' a place and 'where the action is' are everyday expressions reflecting the importance of activity. Activity is movement and goings-on; it is one person walking, or passers-by gossiping on the street corner. Two people regularly together in space is the presence of place ballet, but one that is very weak. The more people regularly in place, the more active the place ballet – at least to a point. Larger numbers of people foster greater activity, which in turn draws additional people and activity.

Place ballets can become too attractive, in the sense that activity outstrips capacity for support. Crowding, congestion, poor service, lack of goods or similar problems then occur in various combinations. The place ballet may be destroyed; alternately, some participants may no longer come, returning the place to a level of activity in harmony with its supportive level.

Regularity fosters attraction. Place ballet organises expectations in time and space; it turns over to body-subject repetitive, day-to-day needs like buying food, eating lunch, getting a postage stamp, doing laundry, walking home. Regularity is an integral part of Relph's existential insideness, whereby a place extends at-homeness and allows its participants to feel profoundly inside (1976b, p.55). Regularity, in other words, fosters attraction because people feel a need for continuity and stability in terms of place and time. Place ballet can fulfil such a need.

Physical design most directly promotes attraction when it works to make a place appealing in some way – for example, aesthetically, or in terms of physical convenience. Because a key to place ballet is prereflective regularity, however, such direct attempts may be superficial and have only fleeting impact. Design has greater influence on attraction if it develops a physical environment that intermingles different activities and the movements required to get to them. This mixing brings participants together spatially and fosters a sense of interaction, motion and life.

Two techniques are valuable here. *Channelling* manipulates routes and pathways so that people's movements intermingle. Many entrances to a building are made fewer, for example, so that people entering will more likely meet. Open edges of a park are planted with shrubbery so that the space can be entered only at certain points along the edges rather than anywhere as was the case in the past. As a result, users are more likely to meet one another. Channelling works to foster face-to-face meetings as people move along in space.

Centring establishes some kind of centre to which many people come – for example, centralised mailboxes, laundry or dining quarters; a market square; a fountain in the middle of a park. 'The finest centres', says Jacobs (ibid., p.195), 'are stage settings for people.' Simple requirements – for food, water, washing, paying bills, sitting after supper – can be valuable activities around which to promote centring. How many place ballets in traditional and primitive cultures arose around the watering place? Even in spite of its size, how often at the supermarket do we meet people whom we would never expect to see otherwise?

Diversity

The crux of attraction is often *diversity* – the place provides several different reasons for participants using a place. People feel foolish going to, or don't go to, places for which there is no reason to go. The more reasons to go to a place, the more often the user will get to it. This effect is multiplied many times by many users. Consider, for example, a successful student-faculty lounge. Usually it sustains a variety of functions: an environment in which to relax, a space through which one must pass to reach secretaries; a place to get coffee, to sharpen a pencil, to store lunch, to get one's mail; the home of Xerox and collating machines. These different uses are more or less regular, yet they provide many reasons to come to the lounge; as a result, people interest in space. The lounge gains the added reputation as 'a place to meet people'.

The greater the place's diversity, in general, the more participants using it. In the case of streets and districts, for example, Jacobs (1961, p.150) argues that diversity ensures 'the presence of people who go outdoors on different schedules and are in the place for different purposes, but who are able to use many facilities in common'. Uses which in themselves are anchorages and draw people to a place Jacobs (ibid., p.161) calls *primary diversity.* Residences, offices, factories, certain places of education and entertainment reflect primary diversity at the neighbourhood level; relaxing, mail delivery, coffee-making, room as pathway to secretaries reflect primary diversity for the lounge.

In addition, place ballet often houses what Jacobs (ibid., p.162) calls *secondary diversity* – functions or enterprises that develop in response to the presence of primary diversity. Restaurants, luncheonettes, book stores and craft shops are examples of secondary diversity at street scale. The lounge as a place to meet people or to poll them reflects secondary diversity in interior space.

In terms of physical design, diversity is helped by a mingling of mixed uses. Users of one activity pass by other activities that they might not come close to otherwise. A synergy is created whereby the individual function shares in the presence of people going to other functions, thereby gaining a greater flow of potential users than if it were spatially alone. Jacobs (ibid.) has emphasised the significance of such mingling in fostering the success of place ballets in streets and districts – at least for larger cities. She also emphasises the importance of three other factors: small blocks; a range of buildings varying in age and condition; a sufficiently dense concentration of people (ibid., pp.178-221). How these four factors are present in particular street ballets and how they might have parallels in place ballets at other environmental scales are important questions worthy of further phenomenological research.

Comfortableness

Comfortableness relates to bodily and psychological comfort and convenience. It involves freedom of movement and activity in a place rather than infringement and delay due to external devices or regulations such as elevators, locks and keys, stoplights, checking-in, speaking to the doorman for permission to enter. Place ballet succeeds best when there is ease and flow of movement.

People using a place normally like to move freely at a pace that reciprocates their reasons for coming to that place. Waiting for an elevator, checking in one's parcels before one can enter a store, calling on a telephone to reach a fellow worker in the same building – these situations involve artificial stops and starts that interfere with the natural flow of a particular activity. Too many artificial interruptions impede the flow of place ballet.

The irritation that drivers and pedestrians often feel when waiting for a stoplight points to the crux of comfortableness. Movement is normally one smooth single process. One stops in transit only because one has reason to. Devices or rituals that regulate movement may be invaluable for reducing congestion or providing security, but experientially they interfere with the natural pattern of movement: to go in one smooth flow.

Comfortableness, therefore, is related to naturalness of experience and human scale. Place ballet will generally be less successful in those places that atomise behaviours or make the body dependent on technological devices beyond its own locomotive powers. Skyscrapers, shopping malls, supermarkets and other centralised, massive environments are all guilty of uncomfortableness to varying degrees because the body alone cannot get around in them but must rely on external back-ups such as an escalator or a system of loudspeakers. In addition, these places often foster uncomfortableness because they set up various artificial obstacles – receptionists, security checks, doors that can be opened only with a special magnetic card.

Comfortableness requires a scale suited to the human body and its reach independent of mechanical extensions – for example, upper storeys that can be reached by walking, enterprises close enough to each other to allow users to move between them easily and quickly, interiors constructed in such a way that people can readily determine where other people are in the space. Comfortableness arises most naturally in places and buildings constructed by human hands without the assistance of sophisticated technological equipment that builds larger-scaled environments than man could by his own bodily devices. In the same way, individual persons and groups who live in and use and own the places they build are more likely to create comfortable environments than impersonal organisations such as corporate enterprises or governmental agencies that normally build more to take advantage of economies of scale or needs of centralisation.

Distinctiveness

Distinctiveness refers to those qualities of place ballet that give it a sense of identity – a sense of being a distinct entity in the midst of a larger environment. 'Atmosphere', 'character', 'sense of place' and similar phrases all capture the essence of distinctiveness.

Distinctiveness may in part relate to physical characteristics of place – clearly marked boundaries, cobblestone streets, a canal passing through its centre, stylish decor, a peculiar location or unusual landmark. More significantly, distinctiveness is related to the people and activities of place. They generate a special regularity, dynamism and atmosphere that attract new users and bring back regular participants over and over again.

Distinctiveness normally develops because originally a place is a taken-for-granted context of someone's daily living. If for whatever reason place ballet gains an image of distinctiveness, outsiders visit and perhaps they themselves become regular participants. The 'distinctiveness' of an ethnic neighbourhood, for example, is the atmosphere of at-homeness for the larger number of participants; it is an explicit object of attention only for tourists and other outsiders. Place ballet may begin to die when outsiders outnumber the people who feel 'inside' that place. Alternately, the place may become a reproduction of itself – for example, the French Quarter of New Orleans – and creates a place ballet founded largely on tourism.

Authentic distinctiveness develops organically, in its own way and time. The constellation of people and events in the particular place nurture a unique living whole which over time takes on its own character and becomes a place and name in users' cognitive consciousness. In contrast is *inauthentic distinctiveness* – places that at heart are the same but which are made to seem different, usually for economic gain. Examples are the different-coloured and different-textured façades of suburban housing tracts, each trying to make itself distinct; or the essentially standardised and homogenised enterprises along commercial roadside strips – the doughnut shop with the huge plastic doughnut, or the steak house modelled after a Western cow town.

Successful place ballets normally generate a distinctiveness which is authentic and natural – unique because of the special combination of insiders, outsiders and events associated with that place. Inauthentic distinctiveness may draw the unsuspecting and unwary to it, but typically the ruse works only temporarily. In time the place survives or succumbs according to its essential qualities as a place and its relationship to the larger environment in which it finds itself.

Distinctiveness arises organically and can not be directly planned or controlled. Perhaps the best approach is to provide monies and policies whereby insiders can work to strengthen their place in their own way and time. Ideally, plans of action should be developed and carried out by the insiders themselves. This process increases the insiders' involvement with place and enhances care and concern.

Invitation

Successful place ballets *invite* would-be participants in. The place projects a sense of insideness. The potential user finds himself in the role of outsider, but because of a sense of invitation, he considers entering the place ballet and partaking of its parts. Invitation has a major role in place ballet. If the person feels invited in, he will return, perhaps again and again. The stronger the sense of invitation, the stronger the attractiveness of the place ballet. The reverse is also true.

Distinctiveness is one element fostering invitation. The place ballet projects insideness because it is a distinct whole separate from surroundings. The would-be user learns of the distinctiveness, directly or second-hand. He develops a wish to visit the place. The nature of the boundaries of place ballet also have much to do with invitation. Permeable boundaries allow the outsider to enter easily. They draw him in freely and sometimes he knows that he has entered only after he has arrived. Open doors, many windows, crossable streets, sounds from within, a vista sweeping into place – aspects such as these mark out a permeable edge allowing easy entrance. Closed doors; wide streets filled with speeding cars; blank façades; no visual, aural or olfactory allurements – these features shut the person out or insulate him from place.

Technological devices often have a major role in making the edges of place ballets less crossable. Traffic, expressways, long high walls of huge buildings all create edges that the person can not cross or crosses only with considerable effort. Air conditioning is another example. Its use requires closed doors and screens replaced by windows; passers-by on the street can less likely sense events inside shops and other establishments. Inside and outside are less connected, and the outsider feels less of a pull of invitation within. People inside the place perhaps feel more comfortable, but this comfort arises at the expense of the larger place as a whole.

Attachment

Attachment is the sense of responsibility and devotion that participants feel for place ballet. Attachment is intimately related to at-homeness and dwelling; it incorporates sparing and preserving, care and concern. The automatic impulse to remove a piece of litter from the pavement or to tell the luncheonette cashier honestly what one just had for breakfast – these authentic gestures arise spontaneously out of attachment. Users who feel attachment will instinctively care for place. They feel a stake in it; the place is an extension of their own selves. At the same time, they feel more human, experiencing a concern and obligation beyond their own personal needs. They feel a part in a larger human and place whole.

People generally can be attached only to humanly scaled places. To feel care and responsibility for a complete shopping mall or a skyscraper or an airport terminal is next to impossible. Attachment is also founded in ownership, either private or collective. Owners feel regard for their place because it is a part of who they are. Users recognise this regard and return it in kind. The placeless environments owned by nameless stockholders or the anonymous public have no or few individual people at their heart. These places have no one to care for them; they often are vandalised, desecrated, poorly constructed, and poorly repaired or cleaned. Sometimes, as in the case of public housing, these places are actively hated and foment in their users a feeling of helplessness and alienation.

As attachment withers, so does the power of place ballet. People use the place for practical, self-serving ends only. Delays, breakdowns, faulty merchandise, poor service are more common and less humanly dealt with. Arguments or dissatisfactions are less likely to be settled face to face. Rather, settlement comes through the courts or some other impersonal context. The corner-store merchant, for example, is sued by a customer who broke her leg on an ice-covered pavement poorly cleared by a neighbourhood boy who didn't care. If the place fostered attachment, the woman might have accepted the accident gracefully, or the boy might have taken more pride in his work. Loss of attachment leads to a vicious circle: fewer people feel care or responsibility, which leads to even further erosion of these feelings. The unplanned and fragile order of the place ballet collapses and another place moves towards dull and lifeless space.

Attachment grows in individual hearts. Like other qualities of place ballet, it cannot be made to happen directly. Indirect nurturing, however, can be fostered by activities and events that allow people to become involved actively and experience a sense of sharing – for example, clean-up or gardening programmes, annual parades or other regularly scheduled rituals that provide people with a sense of participation and community continuity.

Fostering Place Ballet

Regularity and variety mark the place ballet. Their balance is a rhythm of place: speeding up and slowing down, crescendos of activity and relative quiet. The particular place involves a unique rhythm, whose tempo changes hourly, weekly and seasonally. The arbitrary breakdown of place ballet into descriptive parts is valuable heuristically but makes place ballet more regimented and precise than it is in practice. 'In real life', writes Jacobs (ibid., p.54) of the street ballet, 'something is always going on, the ballet is never at a halt, but the general effect is peaceful and the general tenor even leisurely.' All place ballets involve a dialectic between predictability and unexpectedness, regularity and surprise, calm and activity.

Place ballet is fragile. Its pattern arises not from conscious planning but from the prereflective union of people usually unaware of the whole they help create. Only when the place ballet is weakened or destroyed do its members normally realise their participation. They are surprised, angry or regretful, but the feelings are too late: place ballet once destroyed is almost impossible to resurrect.

Buttimer (1978) argues that in planning places, two complementary views must be considered: the place in itself as a lifeworld for residents and users (the *insider's view*); the place beyond itself as it has links with the wider region and socio-economic milieu (the *outsider's view*). In the past, Buttimer suggests, formal planning has too often emphasised the outsider's view, which looks at places 'from an abstract sky' and 'reads the texts of landscapes and overt behaviors in the picture language of maps and models' (ibid., pp.20-1). At the same time, insiders often become concerned only with issues that affect their particular place; they forget the wider socio-economic context of which their place is a part. The need, says Buttimer (ibid., p.19), is 'a dialogue between those who live in places and those who wish to plan for them'. Perhaps the greatest challenge is pedagogical and involves 'calling to conscious awareness those taken-for-granted ideas and practices within each [of the two views] and then to reach beyond them toward a more reasonable and mutually respectful dialogue' (ibid., p.21).

In nurturing such a dialogue, place ballet could well be important. Insiders come to discover one taken-for-granted component of the place in which they live. They recognise through their own experience the value of place ballets; they feel a wish to sustain existing place ballets and foster new ones. At the same time, outsiders recognise the ballets of places under their jurisdiction; they institute plans and policies to protect place ballets and integrate their dynamics into larger environmental wholes.

Ultimately, the continued existence of places is in the hands of their day-to-day participants. 'The crucial philosophical and pragmatic problem lies in what potential role is allowed for residents to have any creative say in designing [places]' (Buttimer, ibid., p.29). An understanding of place ballet is one way in which insiders might take a more effective role in making their environment a place. They begin to recognise the inherent order of people-in-place and strive to create a lifeworld which supports a satisfying human existence grounded in a liveable environment.

Notes

1. In her description of the Hudson Street ballet, Jacobs (1961, p.51) provides one indication of how place ballets intermingle and fuse: businessmen and women, living on Hudson Street, are picked up by 'stopping taxis which have miraculously appeared at the right moment, for the taxis are part of a wider morning ritual: having dropped passengers from midtown in the downtown financial district, they are now bringing downtowners up to midtown'. One indication of place ballet at a regional scale is Skinner's study (1964) of rural marketing in China. He explains how the regular patterns of itinerant merchants moving from one place to another generate a regularity of buyers and sellers meeting in terms of place – the market town.

2. Many of the ideas developed in this chapter owe their origin to Jacobs (1961). Though she says that her argument is appropriate only for large cities, many of her points seem applicable to place ballets in other settings and at other environmental scales. One of the first aims of the student interested in place ballet should be a careful study and thorough mastery of Jacobs's book.

20 AN EDUCATION OF UNDERSTANDING: EVALUATING THE ENVIRONMENTAL EXPERIENCE GROUPS

The important thing
Is to pull yourself up by your own hair,
To turn yourself inside out,
And see the whole world with new eyes – Peter Weiss (1966, p.46).

Place ballets of the past arose from the day-to-day strivings of common man immersed in the triad of habituality. Place ballets were a taken-for-granted part of the lifeworld; they developed naturally, with no need for foresight. Today, place ballets grow spontaneously less often. They need to be explored and understood explicitly in a way similar to efforts of the preceding chapter.

In making the lifeworld an object of attention, the student moves into the triad of openness (Figure 18.2). He or she works to extend and heighten his or her encounters with the world. The triad of openness is concerned with more sensitive looking and more thoughtful understanding. The person, for example, who sees a place in terms of time-space routines and body ballets, who can ask how that place might have a better place ballet, is looking more carefully at his day-to-day world. The place for this person is no longer a taken-for-granted context of daily life, it becomes a thing and process in its own right and the person may begin to feel concern and responsibility for it.

The environmental experience groups are one way of fostering encounters in the triad of openness. The primary aim of the groups was phenomenological – that is, they served as a vehicle through which to explore essential dimensions of everyday environmental experience. At the same time, however, many of the group members felt a sense of increased understanding (Appendix B). Reports describing the group process emphasised its significance in two ways: as an attunement to unsuspected aspects of the individual's own lifeworld; as a tool for evaluating theories and concepts in conventional social science and education.

Attunement to Lifeworld

The lifeworld is the daily world of taken-for-grantedness. Immersed in the natural attitude, people forget that existence might be otherwise. They live inauthentically in that they accept a world of surfaces and normally never look beneath. The person striving to enter the triad of openness works to penetrate the accepted surfaces; he strives to discover more about himself and the world he assumes. The environmental experience groups assisted people with this aim. Group members began to understand aspects of their own lifeworld.

The process of attunement is expressed in various ways. One group member spoke of an opening and need for sharing – especially with children:

> A whole new world is opening for me in terms of places and situations – my home, my school, my need for people and privacy. I appreciate these things with a deeper understanding of their necessary role in my life. . . The thought of helping other people understand these things – especially children – brings purposive meaning to me (commentary 1).

A second group member felt that the group process was 'one of the most important experiences' of her academic career because it had provided her with new insight into the nature of her day-to-day experiences and behaviours (commentary 4). 'Self-knowledge is a kind of understanding I value highly,' she explained. Because of the group experience, she came to accept the possibility that there may be certain essential patterns of human experience, for example, the significance of centring and habit.

Other reports also highlighted the value of the groups in fostering self-knowledge. 'The group work has carried over and become part of my everyday experience' (commentary 5); 'I'm thinking about things differently all the time' (commentary 12), 'the newly gained level of knowledge that now guides my view of environmental experience' (commentary 7). One group member summarised the sensitising powers of the group in particular depth, and I quote the account at length. Note especially the person's better understanding of encounter:

> Glimpsing completely unexpected, very basic forces that shape my behaviour is exciting. I see a little more fully that I'm not at all in control of many of my everyday movements and feel that this is important to my own experience in many different ways – like understanding a hard-to-break habit like nail-chewing, or working on a new piano piece and being able to tackle difficult passages more effectively. Discoveries from the group were useful this past summer

when I had to adjust to a new living situation. I was able to recognise the importance of routine, of centre, of familiar paths, and being able to encourage their development. . .

What was especially interesting to me this summer was the phenomenon of obliviousness. Catching myself (mostly afterwards) so many times caught up inside with some aspect of a situation so that I forgot about the outside world and felt afterwards as if someone else had had the experience. I observed that sometimes the obliviousness had to do with a goal I'd set. For example, one day I took a hike up Pelican Canyon, with ambitious plans to take another trail down a different canyon on the way back. With my day all planned out, the actual hike became almost automatic – an empty gesture. I caught myself at one point completely preoccupied with random thoughts. The whole tenor of walking the trail changed for a few minutes, as if I'd only just arrived, aware of surroundings that I'd been blind to moments before.

It was a frustrating day, with most of it spent rushing to the next landmark. I see this obliviousness orientation in so much I do and often wonder if there's some way to keep myself more in touch with the moments at hand. Here is where the group work has been helpful because in the past I wasn't even aware of this obliviousness – at least now I see it and perhaps in time I can find more ways of getting beyond it, of actually seeing and looking at the things that are there with me at the moment (commentary 3).[1]

A Tool for Evaluation

Besides fostering a deeper awareness of daily environmental experience, the groups gave some members a better insight into traditional approaches of social science and education. People could better evaluate their accuracy and value. One group member, rereading a paper she had written on territoriality, realised because of the group process that the territorial model she had hypothesised was probably erroneous (commentary 11). Before, she had assumed that 'personal scale and national scale were the same', now she believes that there are important differences:

I know a lot more about personal scale now. I don't see how the idea of personal territories could work on a national scale. Some of the needs are the same if you blow them up far enough, but I don't think the comparison gets you too far – as I had thought before. The group has helped me to see that there are different things going on in the two.

Other members valued the group process because it pointed to a valuable mode of education – a way of learning founded on experiential rather than intellectual knowing. One group member, for example, explained that she came to 'enjoy going to the group meetings and sharing experiences, reacting to and getting reactions from people who I feel understand my interest and enthusiasm' (commentary 5). She became aware of 'a special rapport in the groups probably because we shared experiences rather than intellectual knowledge or ability like most college classes'.

Another group member felt that the experience had shown her a way of study that she could respect because it itself respected the things it studied; the group approach overcame a gap she had felt between academic knowledge and other ways of learning:

> One of the most important things about the work we've done for me is that it has shown me a way of knowing that I can respect. In catching phenomena, not manipulating them, and sharing these observations with each other, I feel as if we've been receptive and open to the thing itself. So many of my friends feel a gap between what they're exposed to academically and the rest of their (more relevant) learning experiences. For me, the work we've done in the group resolves this problem. I don't feel a split; learning about myself has become a part of my education. . .because of the groups. That is the strongest force which keeps me coming back to school (commentary 3).

Meanings for Education

In a critique of contemporary higher education, Grange (1974) argues that the typical liberal arts curriculum today emphasises a mode of education founded in *explanation.* Explanation, says Grange, seeks to understand the genesis of an occurrence; it is 'a process of finding out *why* something happens' (ibid., p.362). Grange believes that the emphasis on explanation has fostered a growing malaise in contemporary education. Explanation leads to 'a curriculum emphasizing methodologies of inquiry' which become 'instruments, tools for the future control of history' (ibid., p.361). The most damaging impact of explanation, says Grange, is the student's losing sight of a particular idea, theory, or method's meaning in relation to his own life and the lives of others. Knowledge founded on explanation is 'turned on' and 'turned off'. The student makes use of his learning in the classroom, as he performs as researcher, or as he works as urban planner. Much of the time, however, as he lives his daily life, his knowledge is forgotten about or drawn from only on occasion. The result, Grange writes (ibid., p.362) is that

> the seamless garment of lived human experience is rent by the very process of learning itself. We do not live as *at times* a mind equipped with the power to know; nor do we live as *at times* an agent whose will swings into action to secure certain results. . . We are humans *being.* The texture, fabric, and 'feel' of our being is always a unified zone of awareness, albeit with alternating and shifting currents of self-understanding. The besetting sin of the liberal arts curriculum lies in its indifference to the primacy of lived experience.

Understanding, says Grange, is the need in education today. Understanding is the coming to see more deeply and respectfully the essential nature of human experience and the world in which it unfolds. The subject of understanding is the everyday world met afresh; that world takes on new and richer facets of meaning which speak to the person's own individual life. Unlike explanation, understanding does not seek the causes of events nor does it predict and control the future. Rather, understanding seeks the *meaning* of events; it helps the person to see more intimately and lucidly the pattern of his own existence and thereby *live* better in the future. Says Grange (ibid., p.362):

> Understanding. . .is directly related to the way in which we actually live. Our human existence is constituted by layers of meaning and these layers, in effect, create the 'world' in which we live. As we perceive meaning so we act. Understanding is, therefore, the human process of *standing-under* our 'world' so that we support it, sustain it, dwell in it, develop it, and articulate it.

In Grange's terms, the environmental experience groups and Goethe's delicate empiricism are both educational tools fostering understanding. These approaches reveal meanings to the student founded in his own experience – of what it means to dwell geographically, of what a particular thing in nature is. Understanding joins the student intimately with what he studies; it becomes a partner, a friend – a thing he wishes to know more about because it tells him more about himself. Understanding places the student in the triad of openness – encounters become more lasting and intense. An understanding of environmental experience, for example, indicates through actual experience the importance of body-subject, the value of routine, the relationship between movement and rest. Such understanding shows the student the way people dwell on earth. It gives him clues to better and worse dwelling. These clues may help him to improve his own life-style and help him to understand and improve the life-style of others. Understanding fosters meanings at the centre of a circle – the self – out from which extent reverberations of relationship to other people and places.

If understanding could be fostered in environmental education, the most significant impact might well be a first-hand recognition of modern man's placelessness, homelessness and frequent disregard for nature. Understanding of these threatening forces, particularly as they undermine ecological and human communities, might provoke individual and group action for practical repair. The landscape and physical environment might be made to reflect a better harmony among people, place and nature.

Note

1. This commentary was written in the autumn following the environmental experience groups, after the commentator had gone to Utah for the summer to work as a Forest Service guide at a national forest there.

21 BEHAVIOURAL GEOGRAPHY, PHENOMENOLOGY AND ENVIRONMENTAL EXPERIENCE

> Once we see our place, our part of the world, as *surrounding* us, we have already made a profound division between it and ourselves. We have given up the understanding – dropped it out of our language, and so out of our thought – that we and our world create one another, depend on one another, are literally part of one another; that our land passes in and out of our bodies just as our bodies pass in and out of our land; that as we and our land are part of one another, so all who are living as neighbors here, human and plant and animal, are part of one another, and so cannot possibly flourish alone; that, therefore, our culture must be our response to our place, our culture and our place are images of each other and inseparable from each other, and so neither can be better than the other – Wendell Berry (1977, p.22).

Much of social science in the last several decades has been premised on *positivism* – the philosophical stance that genuine knowledge is based on natural phenomena and their relations as verified by the empirical sciences. A positivist perspective assumes that 'man and nature may be understood in virtually the same terms' (Samuels, 1971, p.81). Researchers have therefore followed a path marked by the natural sciences and often accept without question that 'what is real is necessarily objective, quantitative, and law-abiding' (ibid., p.81).[1]

In translating this positivist view into procedure, geographers have emphasised the tangible, publicly verifiable aspects of people's relationship with environment and space; they have studied the visible patterns, processes and flows of man's imprint on the earth. Behavioural geography and environmental psychology, although they have shifted emphasis to people's *inner* worlds, have generally accepted the same positivist stance, and developed various methodologies to convert the ambiguity of inner psychological processes into empirically measurable images, attitudes, territories, or some similar hypothesised construct that can be elicited and correlated in ordered matrix form.[2]

The problem with a positivist approach, says Samuels (ibid.), is not so much that man can *not* be objectified; clearly, contemporary research in social science demonstrates that he can. The problem is that 'a science of man that must of necessity objectify its subject can not deal with the whole man, only with fragments' (ibid., p.97). Positivist science can investigate only the empirically discernible, objective parts of human behaviour and experience. The less visible, more subtle portions of human existence – at-homeness, habit, modes of encounter, dwelling – are ignored or reduced to recordable manifestations.

Behavioural geography and environmental psychology, in other words, have frequently fragmented and objectified man's inner situation. Research has generally arbitrarily focused on one small band of experience – cognitive map, territorial defence, a one-dimensional form of encounter – which is represented and explained in some measurable, reproducible example. The best example is probably the assumption that behaviour is a function of cognition. The person in this cognitive approach is not a full human being. He is reduced to a machine-like brain intercepting standardised perceptual input. Through a cybernetic process involving a chain of step-wise decisions, this brain 'acquires, codes, stores, recalls, and manipulates information' about the environment.[3]

The students who use the cognitive approach generally justify its implied image of man because it provides one pathway to explanation and prediction. Further, these students might argue that the approach can readily encompass additional aspects of environmental behaviour and experience – for example, affective knowledge about place or varying intensities of perception and information processing.

Yet the chief limitation of the cognitive approach and other behavioural approaches like it lies deeper and cannot be repaired by piecemeal, in-house manipulations. The essential flaw of these conventional approaches is their *implicit separation of person from his world.* By speaking of a person-environment relationship – even as medicated by inner processes – these approaches define the geographical world as an entity apart from man. The environment in this perspective, says Samuels (ibid., p.59),

> is first something *outside, other than,* or *external* to that which it surrounds and has an impact upon. The world as environment is independent of its subject. It is 'objective' as a world of things separated from or external to other things (italics in original).

Granted, the world *is* an objective entity in a cognitive mode of experience: getting around in an unfamiliar environment is one demonstration of this fact. Much more often, however, *we are the world – we are subsumed in the world* like a fish is joined with water. For most moments of daily living, we do not experience the world as an object – as a thing and stuff separate from us. Rather, we *interpenetrate* that world, are *fused with it* through an invisible, web-like presence woven of the threads of body and feelings. For each person and culture this netting is unique – extending out and joining with the different paths, places, routines and situations which in sum combine to make the particular world in which the individual lives. Underlying all these unique worlds, however, no matter how apparently different on the surface, are the experiential patterns of body-subject, feeling-subject, at-homeness and encounter. Through these generally unnoticed, matter-of-fact processes comprising our human nature, we live *inescapably* in the world. We can repeat Relph's dictum that 'people are their place and a place is its people, and however readily they may be separated in conceptual terms, in experience they are not easily differentiated' (1976b, p.34).

The individual person, distinct from the world, has his or her particular integrity, to be sure. 'In one respect', writes the ecologist Paul Shepard (1969, p.2), 'the self is an arrangement of organs, feelings, and thoughts – a "me" – surrounded by a hard body boundary: skin, clothes, insular habits.' Yet there must be a complementary side of man, Shepard goes on to say – a side which he calls 'relatedness of self'. This view requires

> a kind of vision across boundaries. The epidermis of the skin is ecologically like a pond surface or a forest soil, not a shell so much as a delicate interpenetration. It reveals the self as ennobled and extended. . .as part of the landscape and the ecosystem (ibid., p.2).

The 'vision across boundaries' must be a focus of academic attention in the years ahead. The aim is a deepening understanding of *immersion-in-world.* In his notion of 'relatedness of self', Shepard as an ecologist speaks largely of tangible, biological bonds among people, their environment and other organic life. A phenomenological perspective carries the interpenetration of people and environment further. It demonstrates that experientially, too, man can be regarded not as a shell but as a pond surface. As well as the more visible ecological bonds which link people inescapably with environment, human interpenetration with world incorporates less visible, experiential bonds.

This interpenetration is difficult to record because it flows through the prereflective forces of body and feelings. But it *is* present in human existence and must be considered if the aim is a complete picture of people in their world.

If we begin to understand this interpenetration, particularly in our own individual existences, we can better realise that we are not apart from earth but an integral part of it. We are literally immersed in our geographical world, and this immersion is the primal core of dwelling. To understand the earth as the dwelling place of man, we must understand this primal core. At the same time, we may better understand ourselves and cultivate more care and respect for the world in which we live. In this way, we grow as persons and feel a deepening concern and compassion for ourselves, our fellow men and women, and the earth as our home.

Notes

1. 'Positivism is the view of the world that equates being with what positive natural science can know. To be real for it, a thing has to be perceptible in time and space. The tangibility of things proves their reality. Nothing, in this view, is otherwise than an object. *Being and objective being are one.*' (Karl Jaspers, *Philosophy*, vol.1, cited in Samuels, 1971, p.80, n.124, italics in original). Perhaps the most articulate argument for positivism in geography is David Harvey's *Explanation in Geography* (1969). He writes: 'there is every reason to expect scientific laws to be formulated in all areas of geographic research, and there is absolutely no justification for the view that laws can not be formulated in human geography because of the complexity and waywardness of the subject matter' (ibid., p.169). For critiques of the positivist view in geography, see Guelke, 1971; Samuels, 1971; Gregory, 1978.

2. The range of these positivist techniques is well illustrated in Moore and Golledge (eds.), 1976.

3. I take these words from Downs and Stea's definition of cognitive mapping (1973, p.xiv): 'cognitive mapping is a construct which encompasses those cognitive process which enable people to acquire, code, store, recall, and manipulate information about the nature of their spatial environment.'

APPENDIX A: SELECTED OBSERVATIONS FROM CLARK ENVIRONMENTAL EXPERIENCE GROUPS (September 1974–May 1975)

This appendix includes all observations referred to in the text as well as some additional reports not referred to directly but which had bearing on the preceding interpretation. These observations have been transcribed from tape recordings of group sessions. Changes have been made to improve flow of text. Additional observations can be found in Seamon, 1977.

1 Observations on Movement

1.1 Several group reports indicate the routine, automatic nature of many daily movements.

1.1.1 In the summer I used to come to campus from my apartment the same way every day without fail. I didn't have to think about it – I just did it.

1.1.2 When I go somewhere, I always want to go the same old rote way.

1.1.3 I always go to the library one way and back the other.

1.1.4 It's interesting about the library. I always go the side entrance – not the back one. I look at the other entrance sometimes and say, 'Can you get in that way, can you get to the stairs that way?' But I'm too lazy – I've never found out. I keep going the old way.

1.1.5 I tried to get a friend who works with me on grounds crew to go a different way home from work than he usually does. He lives on Gates Street and usually after work he cuts over in front of Atwood Hall. But one night after work I said, 'Come on, let's take a left and go down to Main, because I live on Grand Street and I don't want to go the long way.' But he said, 'No, I want to go across.' I asked why, and he said, 'Because I always do it that way.' He just wouldn't go the new way.

1.1.6 When I was living home and going to school, I couldn't drive to the university directly – I had to go around one way or the other. I once remember becoming vividly aware of the fact that I always went there by one route and back the other – I'd practically always do it. And the funny thing was that I didn't have to tell myself to go there one way and back the other. Something in me would do it automatically; I didn't have much choice in the matter. Of course, there would be some days when I would have to go somewhere besides school first, and I'd take a different route. Otherwise, I went and returned the same streets each time.

1.1.7 We go the same route home to Philadelphia every time. We don't have to think about it and we don't get lost. It just happens, our getting home.

1.1.8 I was driving to the dentist's office and at one stoplight intersection suddenly found myself turning left rather than going straight as I should have done. Just for a moment I was able to observe my actions as they happened – my arms were turning the wheel, heading the car up the street I shouldn't have been going on. They were doing it all by themselves, completely in charge of where I was going. And they did it so fast. The car was halfway through the turn before I came to my senses, realised my mistake and decided how best I could get back on the street I was supposed to be on. At the time of the turn, I was worrying about what the dentist might have to do with my teeth. I wasn't paying attention to where I was going. Of course, usually I *do* turn at that stoplight because I have friends who live up that street and I visit them often.

1.1.9 About a month ago, my room-mate and I switched rooms. Yet when I'm not thinking about it, I'll find myself walking into my old room when in fact I really wanted to go into my new room. It doesn't register with me that I'm headed for the 'wrong' place, but then once I'm there, I'll note my mistake and direct myself to where I should be going. It makes me a bit annoyed that I do it.

1.1.10 I know where the string switches to the lights in my apartment are now. In the kitchen, even in the dark, I walk in, take a few steps, my hand reaches for the string, pulls, and the light is on. The hand knows exactly what to do. It happens fast and effortlessly – I don't have to think about it at all.

1.1.11 We usually keep a clean dish towel under the sink, but sometimes we wash the one that's there and forget to put back a new one. I've noticed quite a few times that as I'm washing dishes I'll bend down for the towel and find it's gone – yet a few minutes later I'll bend down again for it, forgetting that it's not there. It's silly and I have to laugh at myself sometimes. It just happens and it happens so fast that I forget to remember that I've already looked for the towel before.

1.1.12 A few times when using the phone, I've found myself dialling my home number rather than the one I want. My thoughts will be elsewhere and my fingers automatically dial the number they know best. I guess it's because that number is the one I call the most often. I'll suddenly notice what I've done and become aware of what I'm going to dial.

1.1.13 The things I use when I'm working at my desk are all arranged just so. My envelopes and paper are in the top drawer, and things like my stapler and scissors are in the drawer below. The other day I found myself reaching for an envelope automatically – my thoughts were still on the letter I'd written, but somehow my hand had gotten the envelope on its own. I did it effortlessly and unconsciously.

1.2 Sometimes a day-to-day routine is so unnoticed that people can't remember their going to a place or the route they used.

1.2.1 I can't remember which way I go to the library. You go and you don't even know it.

1.2.2 Sometimes for an early class I'll get to the class and wonder how I got there – you do it so mechanically. You don't remember walking there. You get up and go without thinking you know exactly where you have to go and you get there but you don't think about getting there while you're on your way.

1.2.3 You let your legs do it and don't pay any attention to where you're going.

1.3 Difficulties arise when a movement pattern is upset because of a change in the physical environment or because the person is forced to move in a different way.

1.3.1 I get confused in the snack bar where to pay since they've remodelled the place. I seem to automatically move forward to the spot where the cash register was last year, but now the cups and a barrier are there. I have to stop, figure out where I am, and then go to where the cash register is now – which is where the barrier used to be.

1.3.2 One of my classes was moved to a different room last week, but I forgot and found myself going to the old place. They changed the room from Jonas Clark Hall to Esterbrook Hall and all the class walked to the new room together. But the next meeting I walked to Jonas Clark, forgetting that we were now meeting in Esterbrook. Today I checked myself half-way to the old place. I'd had four or five classes in the old place before they changed it.

1.3.3 I was driving out of a used car lot. It's located on a one-way street, but I didn't notice. I started to turn left like I'd normally do when I'm going to go back the way I came. Suddenly I saw this line of cars facing me. I said to myself 'What's wrong here?', saw the problem, and quickly turned the car in the right direction.

1.4 Difficulties arose when members of the groups attempted the 'do-it-yourself' experiment of going to a place by another route than usual. Balking at the task or a vague uncomfortableness resulted. Sometimes the task was forgotten.

1.4.1 I felt I couldn't do the experiment of going a different way. One day I did go to the library a different way than I usually go, but – I don't know what it is, security or whatever – I usually go to places a certain way – I'll always do it. Like this summer to get from my house to where I worked I would cross the street at exactly the same place every day. Once in a great while I would stay on the same side of the street a while longer because I had to stop somewhere, but I'd feel uncomfortable doing it. I don't know why, really. I kept putting the experiment off – I didn't want to do it.

1.4.2 The experiment was an inconvenience for me.

1.4.3 It didn't even feel like I was going to the place I was going to.

1.4.4 It's difficult to go a different way, it feels uncomfortable. Usually from my apartment to campus I walk on a certain side of the street and today I walked on the other side of Main Street. It felt strange. When I was trying to do the experiment, I got lazy. I said, 'No, this time I won't do it.' It was hard, really – just the fact that I got the directions for the experiment a few days ago, but I didn't try it until today. I've been thinking about it a lot, but it's easy to put off to the next time.

1.4.5 I found myself consistently not wanting to do this, saying, 'Why bother?' I didn't feel like going out of my way. It was so easy not to do it and not go out of my way.

1.4.6 At the beginning I enjoyed the experiment. But then it changed and became difficult. I didn't know we were to do it for just one day – I tried to do it for two or three days, and I noticed that I started to fall back to the old patterns. I managed to keep up one new path for three days to a degree, but sometimes I forgot.

1.4.7 I was able to change my route a few times. It was just when I remembered to do it, otherwise I would be half-way before I'd notice it.

1.4.8 When I set out to go a different way I did it only if I took notice to go that way. Otherwise, I'd go the same old way.

1.4.9 If you're with people they won't let you do it. If you say, 'Let's go a different way,' it becomes a monumental thing. They reply,

But this way is shorter.' It's strange how people have set patterns of doing things and if you ask them to do things differently, just for the sake of doing it differently, it upsets their pattern and they won't do it. It's just a gut reaction. It wasn't as if this kid didn't want to walk another way to see what it was like. It's just that his way was shorter, it was *the* way to go, and there was no sense going another way because his was the shortest way.

1.5 Several people mentioned that they often became attached to certain regular routes.

1.5.1 I seem to be attached to some routes. It's not only that I just go that way usually. But if I'm riding with someone and they don't go the way I'd go, I notice myself getting a little annoyed and anxious, asking myself why this person is going the 'wrong' way.

1.5.2 If I'm riding with someone and they don't go the way I would go, I feel a little uncomfortable, and say to myself, 'Well, how come he's going this way? This way isn't the way to go. Why doesn't he take the right way?' It annoys me. Of course I eventually forget about it, but it doesn't seem right somehow when I first notice we're going a different way.

1.5.3 When I was coming back from Boston last night I noticed two places where just for a moment I felt uncomfortable because we weren't going the way I thought we should be going. I was driving, and at the exit on the pike [turnpike] where we get off at Framingham to go the rest of the way on Route 9, I felt myself not wanting to get off, but to continue on the pike. When I'm by myself I would go as far as 495 and get off there rather than use Route 9, which I despise. But the other people with me usually go Route 9, so I took their orders and went that way. Then, we were on Route 9 and I saw the sign for 495 and I wanted to get on that. I didn't because, again, the other people in the car didn't usually go that way.

1.6 Sense of distance is related to use; sizeable distances used regularly become reasonable distances.

1.6.1 I have a friend in my home town who lives about three miles away. Last fall I started to walk over there and at first it was a long walk. After a while, however, I got to like it, and it didn't seem a long walk any more. It got shorter and shorter.

1.6.2 Distances seem further when you think about them in your mind, but when you get to know them by going, they seem closer. Last year we were going to go to a movie at Webster Square – that's only a half mile away, yet I said, 'Wow, that's a long way to walk!' Now that I know a lot of the city because I've gone to places many times with my ice-cream truck, places like Webster Square or downtown seem closer. For instance, Main Street is familiar and friendly now – it used to seem alien to me.

1.6.3 When you have to go somewhere the distance becomes manageable. At another college I went to, the social science and biology buildings were about a third of a mile apart, and when I was a sophomore I had a class in one followed by a class in the other. At first the distance between the two buildings seemed far and I rushed to make it. But over time the trip came to seem natural, and it didn't concern me any more. If people had to walk that distance at Clark, they'd complain at first because they're used to such shorter distances.

1.6.4 When I would visit friends at other schools I would be shocked. I had no energy to walk anywhere because of the difference in scale. At Clark, our space is so limited and I was used to a small scale. It's incredible when you visit a larger campus. You walk and you're struck by the amount of energy that people on that campus use that we don't need at all.

1.7 Movement to unfamiliar places requires greater attention and energy.

1.7.1 When you're going to a place you haven't been to before you have to be constantly awake, making sure you know where you're headed. You're constantly looking, searching out the place you want, checking your instructions for getting there. All that constant watching takes a lot of energy. Once you know how to get to a place, it's much easier. You just go there without having to exert yourself or figuring out where you're going to.

1.7.2 I dislike finding my way to new places. You can't just go there, you have to be 'on your toes', figuring out if you're on the right street, if you've gone past the house you're looking for, if the house you want is on the right or left. For me it's always a nuisance getting to a new place. You're relieved once you've found it.

1.8 Routes are learned through active bodily repetition.

1.8.1 I can remember what an effort it was getting to work the first few days after I moved to where I'm living now. The trip isn't much further than from my old apartment, but it seemed further at first because most of the streets weren't familiar. I had to think about which turns to take. Now that I've been making the trip every day for a few weeks, I can do it even without thinking about it. It's as if the route unfolds in front of me.

1.8.2 When I worked in Washington last summer, I used to pick up a bus near where I was staying in Alexandria and take it straight to the Library of Congress. As I sat on the bus, I'd read the paper or watch the other people. One morning, a visiting friend offered to drive me to work. I couldn't reconstruct the route the bus took – I couldn't give him directions. I could only picture a few of the spots along the route, like the place I got on the bus and the bridge over the Potomac.

1.9 The body houses a sensitivity that manifests in gestures and movements as well as skills and activities.

1.9.1 Last week I was wading in a stream that had a rough rock bottom. I was observing my feet move along. They did the movements all by themselves – I didn't have to do anything. One foot would come down gingerly, feel the bed for sharpness and support. If the place was right, the foot would come down and grip, while at the same time the other foot was releasing and moving forward to find a safe spot on which it could rest. They did this action quickly, and at the same time something in me was balancing, keeping the body erect. At first my attention wasn't on this movement. Then I thought about the group and began noticing how I was moving.

1.9.2 The other day I was walking up the stairs of a new building downtown. These stairs were uncomfortable – my feet had trouble getting in time with their spacing – they just didn't feel right. I noticed them because the week before I had been to the Boston Public Library and we were walking up the stairs there to get to the art galleries. I was struck by the comfort of these stairs. My feet felt at home and moved up them easily, whereas these uncomfortable stairs were difficult to manage – they didn't fit my feet.

1.9.3 The other day I was walking down from the fifth floor of the library to the third and as I came to the fourth floor door, I almost went in, but something stopped me. I continued down to the third floor. At the point that I almost entered, I could feel my body moving ahead – all set to go in. It looked at the door sign, and said 'That's not where I want to go.' But my body wanted to do it – it was flowing on. It was only through some kind of consciousness that I could intervene and do what I wanted.

1.9.4 I went to visit a friend who has a metal shop. I asked him to make a rear spring for my car. I watched him drill a plate and then sand it. His movements were incredible – they flowed together. Both hands were working at once – it was smooth and beautiful. All the time he knew what he was doing and his hands flowed along, doing exactly what they had to do perfectly. He worked like an artist.

1.9.5 I operated an ice-cream truck this past summer. On busy days I'd work as fast as I could, especially if there was another truck near mine. The more people I could serve, the more people would come to my line. As I worked, I'd get into a rhythm of getting ice cream and giving change. My actions would flow, and I'd feel good. I had about twenty kinds of ice cream in my truck. Someone would order, and automatically I would reach for the right container, make what the customer wanted, and take his money. Most of the time I didn't have to think about what I was doing. It all became routine.

1.9.6 I've noticed when I'm playing the piano that, especially with passages that I know fairly well, I play more easily when I let my hands do the playing and pay attention to the dynamics of the piece. I can remember when I was younger and took piano lessons. I would purposefully set my attention on something other than playing – I used to think about bowling – because I had noticed that the playing went more smoothly.

1.9.7 When I started working in a post office last summer, I had the hardest time putting mail into the boxes. The boxes were arranged numerically, and I'd have to stop for each piece of mail, and search for first the proper row and then for the particular box. I couldn't understand how the other workers could do it so quickly. After I'd been working there for a little while, the whole nature of the job changed. I found that my hands reached for the proper box. I didn't have to check the numbers any more. The whole process was one smooth flow. I could let my attention wander and still do the job quickly.

1.10 The body has some ability to adapt creatively to new situations.

1.10.1 I've had to use my friend's car which is automatic shift. I had never driven an automatic before. I've noticed how quickly I adjusted to it. A few times my left foot reached for the clutch, and my hand wanted to shift, but soon they stopped and the process proceeded easily. It was as if nothing had changed.

1.10.2 My car broke down while I was home. I brought my mother's back. It's a big 1968 Chrysler and at first I wasn't sure I could drive it. It's much longer than my small car. Driving the larger car felt strange at first. I didn't know how far its sides extended. I noticed that if I didn't worry about it but let the driving happen – just hand it over to my hands on the wheel – they automatically knew what to do, and the driving was easier. Soon it was as if I'd driven the car all my life.

1.11 Habitual movements and routines extend in time as well as space.

1.11.1 It's possible to know where my grandmother will be at any moment of the day. She is always in a particular place at a particular time and usually doing a particular thing there. Like between six and nine she'll be in the kitchen helping to cook and clean. Then from nine to about twelve she'll be on the front porch sewing.

1.11.2 Unless I'm away or something special comes up, I have an early morning routine that I follow every day but Sunday. I get up at 7.30, make my bed, go to the bathroom, brush my teeth, comb my hair, return to my room, take off my pajamas, dress, check pockets for money and keys, then walk up Main Street to the corner café. There I walk in, pick up the *Times*, sit down – usually in one of the booths further back – and have an order which is usually the same – one scrambled egg and coffee. Then I eat and read the paper. I like this routine. I've noticed that I'm bothered a bit when part of it is upset – for example, if the *Times* has been sold out or if the booths are taken and I have to sit at the counter. It's not that I figure out this schedule each day – it simply unfolds and I follow it.

1.12 Routines happening in a supportive physical environment often foster a wider place dynamic.

1.12.1 Since I've started work, I've found there are customers who come in fairly regularly. One woman comes in almost every night at 5.30 to buy milk, and an older man comes in at six to buy cigarettes. There are several others, and I've got to know them and say 'Hello,' since they're regular faces. I like seeing people I recognise. It helps pass the time and gives me some people to talk to.

1.12.2 One thing I've noticed about the activity in the corner café between eight o'clock and nine is a certain regularity. Several 'regulars' come in during that period, including the undertakers across the street, the telephone repair man and several elderly people, including one woman named Claire, whom I know and say 'Good morning' to each day. Every morning she comes in at about 7.45 after mass at the church across the street. Many of these people know each other. The owner of the place knows every one of the regulars and what they will usually order. This situation of knowing other people – of knowing who's there at the time, recognising faces that you can say hello to – somehow makes the place warmer. It creates a certain atmosphere that wouldn't be if new faces came in every day.

2 Observations on Rest

2.1 Wherever people go, even for short lengths of time, they seem to establish centres.

2.1.1 When I was in New York, I organised the city around my sister's apartment. The first day I did the one thing I had planned and then didn't know what to do. I took a bus back to the apartment where I was staying. I did it without thinking. I could have done anything, but I went back there.

2.1.2 I went to Albany, New York, this past weekend. I quickly got my bearings in terms of the friend's house where I was staying and came and went in terms of it. His house became a centre. It seems as soon as you go to a new place, you immediately establish a centre and move around it. Even if you stay at a place for only one night, it's still a centre in your mind.

2.1.3 Even for a few hours in a place you have a centre. When you stop by the side of the road to eat lunch, and stay for a time resting – even there you pick a place, sit down, and then usually spend the rest of your time in terms of that place. Or when you wait in a bus station for a few hours, you get up for some candy, or go to the bathroom, or take a walk, but then more than likely, you'll return to the same seat, which is the centre around which you organise the bus station.

2.1.4 When I go shopping, especially in a place with which I'm not too familiar, I've noticed how the car becomes my focus in space and I direct my shopping movements in terms of it. I park the car, note its location, go do my errands, and return to the car. While I'm there, the car is a kind of centre that I hold in the back of my mind.

2.2 The centre of a person's lived-space is generally the home.

2.2.1 I don't move all over the earth. For periods of time the space I live in has a kind of centre and that centre is my home. I go to sleep there at night, I get up there in the morning. From it I leave in the morning, do my tasks during the day, and then return in the evening.

2.2.2 Space isn't all equal for me. Where I live is a unique place because I'm always leaving it and coming back. In one sense, I'm bound to that place.

2.3 People become attached to the home.

2.3.1 I'm going to move in with some friends, and I'm feeling bad about leaving my apartment. I feel attached to it and I'm going to miss it. I like the location, the windows that give so much light, and my room. It's sad, leaving.

2.3.2 When we were living in Brill House, a student-residence house before it was torn down, it gave us a lot of grief. In the end, the heating system broke, it was cold, and we were told we could have rooms in the dormitories. Yet no one moved even though we could have had free meals on the meal plan. We stayed. We'd just not go back there during the day. We'd work in the library instead. Then we'd go back, make dinner, clean up, go to sleep. We'd be cold and uncomfortable when we woke up. Yet we just had to stay there. Only when the water was turned off did we finally leave.

2.3.3 I remember when the heating system in my apartment was broken for a few days last winter. Friends invited me to stay at their houses but I didn't go. Like New Year's Eve. The friends I was with told me to stay but I couldn't think of doing it. It didn't seem right staying in their place when my apartment was just a few blocks away. It was cold in the apartment, but I wanted to be home. I remember thinking to myself how irrational I was being – that I wouldn't be comfortable and might get sick. The thoughts had no effect. I found myself returning home with no hesitation whatsoever.

2.3.4 This summer I was returning home from a long trip out west. I had planned to stop for dinner at my sister's, who lives about 150 miles from my home, and whose house is directly off the highway on which I was travelling. By the time I got to her house, it was getting late, and I was feeling this need to drive home. There was this irresistible urge to be in my own place and sleep in my own bed. I drove on and reached home around midnight.

2.3.5 I don't feel as settled into my dorm room as I did living in my old apartment. I think it's because all the dorm rooms are the same. It's difficult to make any personal mark. I don't feel as comfortable there as I might if I could make the place my own. Because of the futuristic ceilings and stucco walls, there's little chance to give the room a personality.

2.4 The home satisfies a variety of needs.

2.4.1 I go there, [to the apartment] to get oriented.

2.4.2 My apartment is the place where I can 'let my hair down'. I can do what I want there, not do what I don't want to do. It's my place, I'm free to be what I am there.

2.4.3 I go back there to get myself together before another class.

2.4.4 It's like when you're sick: the only place you're comfortable in is your own bed at home. If I'm sick and away from home, it's always worse than if I'm sick and at home. When you're sick you don't have the energy to pretend you're something you're not.

2.4.5 Sometimes I go back to get strength. For example, a professor had told me he wasn't satisfied with a paper I had written. I found myself walking back to my apartment just to recuperate. I didn't know what I'd do there, but I knew the apartment would help me to feel better.

2.4.6 My home is where I can best be myself. It helps me get away from over-exposure.

2.4.7 Home is the place where I'm separate from the world. I can rest myself there. At home I control the disturbances that might bother me.

2.4.8 My apartment is my special place where I can do things I like and don't feel bothered or guilty. Reading quietly, sitting with a friend, playing my recorder – all these things I can do anywhere, but somehow they seem best done at home. At home I don't feel ashamed to be miserable. I can go to my room, shut the door, be as ugly as I want. I can be angry with my room-mate and it will be okay. No strings are attached to anything I do at home.

2.5 The house provides privacy; loss or lack of privacy may make the person uncomfortable.

2.5.1 I had a room-mate last year who was very easy to get along with. But he was constantly in the room. It felt like the room wasn't mine because he was there so much. He would do everything possible to accommodate himself to me and I would do the same for him, but it still felt like the room wasn't mine because he was there so much.

2.5.2 Until a few months ago, I had only one room-mate. Now I have two. Before, the apartment seemed mine; I could always come back there some time during the day and have the place to myself. I could be alone and have some privacy. With the additional person, it seems like there's always someone there and I never have the place to myself. It's a relief every so often to know that both room-mates will be away and I can have the apartment to myself.

2.6 Violation of home-space can create tension, even anger; rituals are often established to safeguard the sanctity of home.

2.6.1 A friend has been staying with us until he finds an apartment. Lately he has been entering the apartment without knocking, and I've gotten angry about it. He doesn't live there and he should knock before he comes in. He's a guest, not a resident, and he violates our privacy when he walks in like that.

2.6.2 For the past few weeks, workmen have been renovating the apartment house where I live. I've been trying to observe my reactions to them. There's a feeling of trespass: 'What are those people doing here in my building?' I find their presence annoying. I realise they have to be there, but there's tension having them there. I'm especially anxious when they come into the apartment – like to install a new intercom system. The whole apartment feels different – like you won't feel comfortable until they leave. It's like you can't be who you usually are at home while they're there.

2.6.3 My mother and father lock the door every night before retiring. Usually my father does it, but then my mother re-checks to make sure he hasn't forgotten. My father has a regular routine. He goes to the outer porch door, flips on the yard light, checks the outdoor thermometer, shuts off the light, locks the porch door, then comes in and locks the inner door which comes into the kitchen. Then about fifteen minutes later my mother gets ready for bed, and she checks the door, too. In the morning, my mother gets up first. As soon as she's downstairs, she unlocks both doors and looks out to see what the weather is like. They've done this as long as I can remember.

2.7 Home may mean an atmosphere of warmth.

2.7.1 When you're in a house that hasn't been lived in for a long time, there's a feeling of coldness. It takes time to make the house seem vital again.

2.7.2 We moved to an apartment in New York City that hadn't been lived in for a long time. I didn't want to live there because it felt cold and unused. Once we had cleaned and fixed it up, it felt better.

2.7.3 I notice a different feeling in houses – whether the house has just been emptied or whether it has been vacated for a long time. There's a definite feeling, a lack of energy in a place where no people have lived for a long time. It feels like a ghost town.

2.7.4 In Baltimore, I was living in a row house that looked like all the other houses on the street. Our house looked as if no one cared for it – it was a mess. There was a young kid who lived next door with his parents. He would come over to visit us. We went over to his house once, and there was such a difference. There was a sense of group and caring – the place looked nice. Someone had taken pains with it. It was decorated with family pictures and things from trips. I liked the house. It had a warm feeling, I felt very good being in it.

2.7.5 My parent's house has several rooms which are rarely used, for example, the formal dining room. No one goes in it except when there's company. You walk in and the room feels cold – I don't like to be there.

2.7.6 I have vivid memories of the living room in my grandfather's house. It wasn't fancy or new, but all old things worn and well used were in it. It had a quality of warmness. There was a stuffed deer's head over the mantle, and I remember one time lying on the rug by the fireplace looking up at it. I remember feeling warm and happy, snug and secure.

2.7.7 I remember working for the Census Bureau in Albany, New York. One day I went to an especially dingy house. I expected to find a dingy apartment, but instead I walked into a place that felt warm. A divorcée lived there with her two children. The place felt so much a home. It was decorated in light blue and was clean and ordered and cared for. It felt warm and cosy. I almost wished I was a child living there – it felt so supportive. The place felt so much a home.

2.8 The body houses a knowledge of the home.

2.8.1 I was on my way out of the Geography building, planning to go to the post office and then to the corner café for breakfast. As I was leaving, someone told me that the reserve readings I had supposedly put on reserve were not there. Immediately, I began thinking up all sorts of reasons as to why the 'shiftless' library staff had not done their job. Suddenly I found myself walking up the stairs to my apartment which is right across the way from the Geography building.

2.8.2 Yesterday I helped a friend take a heavy trunk down to the bus station to be shipped. On the way back, we were having a lively conversation and suddenly I said, 'How dumb! Here I am driving us back to my house when I have to take you home.' He said, 'Yeah, I was wondering where you were going.' He lives in a direction completely opposite to my house.

2.9 Routines are associated with the home.

2.9.1 On working days, my father follows the same routine each morning. He automatically gets up at 7 o'clock – he doesn't need an alarm. He puts on some old clothes, goes to the bathroom, then picks up the morning newspaper from the front stoop. He puts two sausages in a pan over low flame. They'll be ready to eat at 8.15. While they cook, he reads the paper, always sitting in the same chair. He slouches. Just before the sausages are done, he softboils an egg; he doesn't even wash the pan but uses the same water day after day. He puts a piece of rye bread in the toaster and pours a glass of orange juice. . . After breakfast – he calls it his 'three-minute breakfast' because that's how long it takes him to eat it – he puts the dishes in the dishwasher, shaves, bathes, dresses, and leaves the house at 8.50 sharp.

2.9.2 My brother routinises the things he does at home. For example, he has a dinner routine. He gets home a little after 6.30, puts his briefcase in the dining room, goes upstairs to change his clothes. Then he makes his dinner – a salad, a bowl of either canned ravioli or spaghetti; he says he doesn't want to make a choice of menu every day. He eats in front of the seven o'clock news on television.

2.10 Social harmony can be an important component of the feeling of home.

2.10.1 When we moved into our apartment, some subletters were living with us, but they weren't friends. They were just helping to pay the rent. When you went home it felt like an apartment, it didn't feel much like a home. As soon as our friends moved in, the place changed. It's nice now. Even when I return and nobody's home, there's a good aura about the place. We all get along really fine, we eat dinner together every night, and it's just like home, it really is. You look forward to eating together. It seems like we've developed a family feeling.

2.10.2 I definitely noticed that in the time I was living with some people I didn't get along with, there was no energy to spend on new activities. One of the reasons I had come back to school was to grow as a person, to try out new things. The uncomfortable apartment situation made me depressed – it upset me inwardly. I had no wish to get involved in anything – just the minimal to keep me going. After the move, when I felt more comfortable again, I had more energy. I was settled and I could give myself to new things. I got involved in a pottery course and began giving volunteer help in a nursing home. I felt that I could involve myself in new things because I felt more free inside.

2.11 A lack of centre may generate distress.

2.11.1 Until a few weeks ago, I was living in an apartment that didn't feel like home. My room-mates were hostile, and I was uncomfortable about being there, about going back there. The place had no drawing power. I felt disoriented because I didn't have a centre anywhere.

2.11.2 Most of the time when I'm travelling, I'm travelling to a place, and I *know* I'm going to a place. But when I travel for a long time and go quickly from one place to another, I find that it can be disturbing, because you don't know exactly where you're going to be.

2.12 The guest-host relationship has bearing on the nature of home.

2.12.1 It's uncomfortable entering a stranger's house because you feel passive and unsure. You don't know where to go, where to sit. You just stand there, look and wait for someone to tell you what to do. It always feels good when your host welcomes you, takes your coat, sits you down, puts you at ease.

2.12.2 I find that most college students are terrible hosts. They seem to expect that you should walk right in, make yourself at home, feel one hundred per cent comfortable. Sometimes people have even left me alone, or left me with people I barely know. I always feel uncomfortable, I feel at a loss, wishing my host would be more of a host.

2.13 There are centres within the home.

2.13.1 My family *always* sits in the same seats at the dinner table. We bought a round table for our house and it seems that it would be easy to frequently switch chairs, to sit wherever you want just because the table is round. But after a while, everyone settled down in his 'own' chair. I remember wanting to sit somewhere else for a change, but everyone else put up such a big fight that I never could. It's the same for classrooms; if professors assigned seats everyone would 'flip out', but lots of times people end up sitting in the same chair every time anyway.

2.13.2 My father has a seat in the living room and the moment he walks in, the person in his seat will say, 'Hey Dad, do you want your seat?'

2.13.3 When I go home, I always automatically grab the chair I used when I was a kid. I pick it up and drag it across the floor to my old spot at the table. But my mother likes me to sit on the other side of the table where my sister used to sit because I'm less in the way over there. So, almost every time, I'll be pulling the chair over and Mom will say, 'Oh, are you going to sit on that side of the table?' And then I'll remember, put the chair back, and go to the other side. I still don't feel comfortable on that side, but my Mom likes me there so I do it.

2.13.4 My desk and the big rocking chair I got from the Salvation Army are the two places where I usually am when I'm in my room, I feel attached to both of them, especially the chair which faces out the window and has an attractive stand-up lamp behind it. I like to read rocking in that chair before I go to bed at night. These are the two places I naturally go to when I walk in my room and I want to sit down.

2.13.5 I always find something not quite right about sitting on someone else's bed unless they definitely say, 'Sit down.'

2.13.6 A good friend of mine was sick in bed today and I sat down naturally next to her. We were talking and just for a moment I thought, 'I'm sitting on her bed – I shouldn't be.'

2.13.7 I didn't have a desk in my room last year and I never had a place to study there. This year, the first thing I did was to make a desk, and it makes a world of difference. I know where I can study now.

2.13.8 While I was home, I noticed how much time my family spends in the kitchen. All the important discussions go on there. It's the same in my apartment – we spend hours around the table talking. If you go out for a while, you're drawn back.

2.13.9 It's getting colder in my apartment and we're all gravitating to the kitchen. There's a heater in the living room, too, but we never sit there. For one thing, there's no table there. The kitchen is the most important room. We did most of the fixing up in the kitchen, and all our plants are there. The room has an atmosphere of friendliness and cheerfulness.

2.14 There are important centres outside the home.

2.14.1 The corner luncheonette is a kind of centre for me. It's a good feeling walking in there.

2.14.2 I like the pizza parlour down the street. They have good food, I can sit down there and feel comfortable. People get loyal to that place. I had a friend who was driving across country and he had to have breakfast there before he set out.

2.14.3 The crafts centre is a very special place for me. A magnetic force draws me there. It seems like I can't get away from campus without going there.

2.14.4 The bakery on Main Street seems near to me – we go there often to buy bread. The people are friendly there.

2.14.5 The nearby park has come to mean a lot to me. I'd feel sad if anything happened to it. I go there for walks to clear my head. I have a definite route I go: to the left of the pond, up the hill to the old stump, then back on the other side of the pond. I go to the same old stump, stop there for a while, sometimes sit down, relax. I feel close to the park. It helps me to get away from the university.

2.14.6 The park across the street from my apartment has become a special place for me. There's enough space there to get away from the city – to be with the trees and the green grass. I like to be there by myself and sit quietly.

2.14.7 I spend a lot of time in my office these days. It offers a good blend of privacy and social contact. Sometimes my office-mates are there, other people come in, and we have a good conversation. At other times I have the place to myself and I get my work done. I wouldn't think of working elsewhere. I feel close to it.

2.14.8 At school, my favourite place is my desk – it seems to be the centre of what I do.

2.15 The body houses a knowledge of these centres.

2.15.1 Coming from my office, which I use a lot, I stopped at the bathroom as I was on my way to a downstairs room to get a map. In the bathroom I was thinking about a class that I had to prepare for and I walked out deep in thought, heading to my office. About half-way there I remembered I hadn't gone to the room I had intended to go to. There was a sudden moment of remembering, a quick bit of annoyance. I turned around, went to get the map.

2.15.2 Since they've moved the snack bar to its new location, I feel a little uncomfortable when I go there. It seems wrong walking to the new place when in the past I've gone to the old location. I still go, but it seems strange. It will take me some time to get used to it.

2.16 Many things used in day-to-day living have a resting place.

2.16.1 When I go to bed, I put my glasses in a definite place. I keep them on my desk right above my head. When I get up I put them on automatically. I'd never think of changing their place.

2.16.2 I always keep change, keys and pens in my right pocket and tissues in the left. When my right pocket gets a hole in it, I have to change things to the left. Then I never know where to look.

2.16.3 I have specific places in my pocket-book for certain things. In the pouch in front I have pencils and pens. Inside is a zipper case where I put my keys.

2.16.4 I have a general area in my room where I put my glasses and notebooks. I find them there when I need them. If I haven't put these things there, I'll often forget them.

2.16.5 My desk top has a series of places where I keep things. Scrap paper is always on the left side, pens and pencils in the upper left part, and books on the right.

2.17 The body houses a knowledge of these places.

2.17.1 My mother knows the exact location of everything in our house. She has a place for everything. She doesn't have to figure out where a particular thing is – she goes to it automatically. I'll need some string, for example, and she'll go immediately to the right drawer. I'd have to check a few places before I'd find it – if I did then.

2.17.2 Because of the group, I've come to be more consciously aware of how important places are to me in the kitchen. All the things I use have definite places – even the spices in my spice rack. When I'm preparing a meal, I can quickly locate ingredients and utensils without having to think about it at all. Everything is at hand and ready for use.

2.18 New places often conflict with old patterns.

2.18.1 When I was younger, we had a clock in our kitchen which was in the same place for six or seven years. It always was above the refrigerator. We moved the clock to the stove. I remember that for maybe some three years after, I would sometimes 'do a double-take' in terms of that clock. I'd look up at the wall to get the time and the clock wouldn't be there. The clock had a new location and still everyone would look for it in the old place.

2.18.2 My father washed the dining-room rug and he had to move the table. Usually it's located directly under the chandelier, but when he moved it back it wasn't quite centred. We were eating dinner and family members began to notice the change. We had to get up and move it to its proper place. My mother noticed it first, but everyone was bothered about it.

2.18.3 Last night I went to use the typewriter in my office. Someone had turned the typing table around so that the typewriter was facing in a different direction. I started typing but it didn't feel right and I turned it back around.

2.19 The do-it-yourself experiment of moving a thing to a different place points to the role of habit in establishing spatial order.

2.19.1 I decided to switch the utensil and silverware drawers around this past week. We asked our other room-mates if it was okay and they agreed. Throughout the week, people mentioned the change frequently. Someone would go to a drawer, open it, realise it was now the wrong one, then go to the other drawer. It got to the point where we would just walk to the place it used to be, remember the change, and go to the other drawer. When I went to the drawer – usually it was for silverware – my mistake wasn't exactly frustrating – it was like we were doing it as a game and so we'd talk about it. I remember people saying, 'Ah, I've done it again!'

2.19.2 I moved my towel to a new place in the bathroom. I'd enter and say, 'I'm not going to let myself fall for the change!' But one day I walked in and was upset and had something on my mind. I walked up to where the towels used to be and I said, 'Oh, no, I've forgotten!'

2.19.3 In doing the experiment, I decided to move my trash-can, which has been in the same place by my desk for two years. It was hard for me to move it – there was this feeling of inertia.

2.20 People may feel a sense of inertia when they consider changing the place of something.

2.20.1 I was thinking about changing furniture around. No matter how illogically a room is arranged, no matter how much sense it would make to change it, I always hesitate to rearrange things. I don't like to do it, there's this drawback feeling that I want to keep it as it is.

2.20.2 I wouldn't want to reorganise my desk, even though it's not arranged in the best way. It would be awkward if it were different.

2.21 People often establish a parking place for their automobile; if they don't, problems may arise.

2.21.1 I sometimes put my car elsewhere than its usual parking place. Later, I'll forget the change and for a minute think the car has been stolen.

2.21.2 I've forgotten several times this past semester where I parked my car. I find myself stopping at the first empty space that looks convenient and parking there. When I go out to find it at the end of the day, I can't remember where I've parked. As I look for it, I find myself thinking, 'Where did I park this morning? Where would have been the most sensible place to park?' Often this logical approach doesn't work and I just have to go around and look. It's ridiculous and annoying at the same time. For the sake of convenience, I'm beginning to establish a parking place.

2.22 Deciding on a place for a thing may sometimes be difficult.

2.22.1 I remember when I was about ten or eleven, our family decided to buy a toolhouse that you could assemble yourself. An argument arose between my mother and father as to where we should put it. My father wanted it by the road so we could get to it easily in the winter-time when there was a lot of snow, but my mother said, 'No, it would look ugly there.' Finally, they agreed that we'd put it at the side of the house.

2.22.2 When my brother got back from Vietnam, he wanted to plant a tree at my family's house. My whole family ended up getting involved with the planting – we took it very seriously. My father thought it should go in the back yard near a tree that he had planted when he first bought the lot 25 years ago. My brother was insistent that it should go in the middle of the front yard, where it would get the most sun and root space, and where it would stand out. In the end, my brother got his way, although he planted it a bit closer to the side of the yard than he had wanted to at first.

3 Observations on Encounter

3.1 Encounter with the world fluctuates and varies in intensity.

3.1.1 It seems that I'm always encountering the world differently. It's almost as if there's not one world in which I live but many. Some days I don't notice a thing – there seem to be plenty of days like that. Other days seem fresh. I don't know what it is but sometimes I'm very close to the world around me and other times distant and non-alert.

3.1.2 It's strange how the world is for me. At times I know I'm not seeing anything. I'm so caught up with my own self inside that the world has no chance to penetrate. Sunday morning, for example, I went for a walk down Main Street. The trees seemed so beautiful and alive. I hadn't seen them that way in a long time. I was feeling good – in a calm and quiet mood. I happened to meet my old girlfriend who said something that really hurt me. I kept walking but now the walk was completely different. I was full of anger inside and didn't notice a thing. It was like a barrier had been put up. My anger and bad thoughts blocked out the possibility of seeing the way I had a few minutes before.

3.2 Much of the time people are oblivious to the external world.

3.2.1 I was walking down a hallway and it took me the longest time to recognise that someone was walking in front of me. I wasn't conscious of his presence, and it surprised me when I noticed him that I hadn't seen before I did. He was there in front of me several seconds, yet it took a bit of time before I consciously realised he was there.

3.2.2 I was in a rush to get here and walked by a friend who was going in the opposite direction. He called or else I wouldn't have seen him.

3.2.3 When I was coming back from home on the bus I was tired. When we got on the turnpike I wasn't sure if we were going to Worcester next or Lowell. I said to myself, 'I'll pay attention to make sure the bus's next stop is Worcester. I don't want to end up in the wrong place.' I forgot, however, and the next thing I knew we were pulling into the Worcester bus station.

3.2.4 We were driving back from vacation on the turnpike and someone said, 'Did we pass the Worcester exit yet?' We had. It was our turn-off, yet everyone had forgotten to look for it.

3.3 In moments of obliviousness, people are engaged inwardly, with thoughts, worries, reveries or bodily discomfort.

3.3.1 I'm always running around, thinking about what I have to do next. As I go I'm thinking about what I just did or where I'm going. In times like these I don't notice many things around me.

3.3.2 When I was driving back to Worcester after intersession I was caught up with thoughts about the week ahead. I was so busy planning out the days ahead that I drove past my exit.

3.3.3 Much of the time this past week I found myself thinking about my problems. There was a lot on my mind and I didn't notice things very often.

3.3.4 I know when I'm not feeling well physically I just look at the ground and try to get where I have to go. There's no energy left to notice things. Being sick 'puts the damper on'.

3.3.5 I often take a walk in the park, but not because I like the pleasant surroundings. It's more because my thinking often improves when I walk. I can figure things out better or work out problems that are bothering me.

3.4 Obliviousness to the world may extend to tasks or pastimes in which the person is engaged at the moment.

3.4.1 When I house-clean I have a routine. One day I was vacuuming, however, and there was one corner which I wasn't sure I had done. I vacuumed it again. In housekeeping it's so easy to 'go off in a daze'.

3.4.2 I've had the problem sometimes of not being able to remember actions that involve repetition. I have to ask myself, 'Have I put enough cans of water in the orange concentrate?' 'Have I added the right amount of water to my two cups of rice?'

3.5 At times people consciously seek to withdraw themselves from the task at hand.

3.5.1 Often when I'm doing something that doesn't require attention – like driving or cleaning – I'll sing to pass the time. Last night I was singing as I was cleaning the dishes, and a friend who was visiting asked why. I said, 'Well, you've got to do something to take your attention off washing up.'

3.5.2 When I'm working at my job as dishwasher I rarely pay attention to what I'm doing. It's easier to daydream or think about what I'll do after work. There isn't very much about the job that holds my attention.

3.6 Even in obliviousness, some part of the person is in contact with the world.

3.6.1 Last week I was walking from my dormitory to the library, lost in thought, making plans for my parents' visit the coming weekend and where we might go for dinner. Just for a few seconds, I was able to watch myself walking up the hill, avoiding puddles that had formed because of the rain we had last week. Something in me was watching the water puddles and guiding me around them. Even though I wasn't consciously aware of each puddle, something in my eyes and feet were working together. I'd see a puddle ahead and my feet would instantly hop over it or go around. All of this was happening as I was immersed in my thoughts. I didn't do a thing; it happened automatically.

3.6.2 Last night I was driving back from Boston, and I was thinking about what I was going to do today – about an appointment I had with a professor and what I was going to tell him about my research paper. I suddenly noticed that I was passing another car. I had done the action without any thought – some part of me was noticing and co-ordinating the road with the movement required to pass. I hadn't consciously decided to pass the car – I was oblivious to driving. Somehow my eyes and my hands on the wheel and my feet on the pedals were taking me safely through the passing manoeuvres.

3.7 Obliviousness to the world may be so complete that an accident results.

3.7.1 I remember once last fall I was walking along Main Street to my apartment, caught up in a decision I had to make about what I would do this semester. I was 'in a fog', listening to all sorts of arguments in my head. I didn't notice the 'No Parking' sign on the edge of the sidewalk, and it caught my shoulder as I walked by. I jarred myself but was surprised more than hurt. It was an unexpected intrusion on my thoughts, a jolt from another world.

3.7.2 Last week I was working my job at food service. I had asked my manager for the next day off, but he refused. I was angry, talking to myself and justifying why I should have time off. Suddenly I heard a large crunch. I had forgotten to keep track of dirty dishes going through the washer. They'd taken up all the available space and wedged together.

3.8 Sometimes people watch the world.

3.8.1 I was sitting on the lawn in front of my dormitory last Tuesday afternoon, watching people, seeing who was going where. I wasn't watching anything or anyone in particular – just looking. It's relaxing. I must have sat there an hour or more, taking in the atmosphere. I'm not saying I was watching the scene the whole time. Sometimes I'd be 'into myself', thinking about things or worrying about school work I should be doing. It was a mixture – lost in thought for a while, then noticing something, on and on.

3.8.2 The other day I was in the park, sitting on a bench, watching the ducks on the pond. They first caught my attention because of the noise and motion they were making. I sat there about ten minutes, watching, seeing what they were doing. It was like watching a movie – they were interesting to watch. Then my attention waned and I left.

3.8.3 Last Sunday I went to the stock-car races and it was one of the most exciting shows I've seen in quite a while. There were three racers scrambling for the lead, and no one had a clear edge. I got really involved – standing, jumping, shouting encouragement. Everyone in the grandstands was up and screaming and waving. It was an exhilarating experience. It was like tumbling back to another world when the race ended.

3.9 People don't necessarily notice the same things.

3.9.1 I was driving with a friend. We went under a bridge on the top of which was a device that measures the speed of cars going under. I said, 'Oh, I'm going seventy-two miles per hour!' 'What made you say that so suddenly?' asked my friend. 'Didn't you see the electric sign?' I said. He hadn't seen it at all, even though the sign was large and visible.

3.9.2 I was driving with a friend yesterday and a car pulled out in front of us. I thought it was an ugly colour and said, 'Ugh, look!' 'What's the matter?' he replied. He hadn't even noticed it.

3.10 Things unnoticed may suddenly be noticed.

3.10.1 I was walking from the drug store with a friend. We walked by an alley that I'd never noticed before. It was something I had never seen, yet I had passed that place many times. I don't know what caused me to notice it.

3.10.2 I was looking down a row of chairs in the library. I noticed that the corners of the two chairs directly in front of me were frayed. I was trying to read at the time. I don't know what drew my attention to them.

3.11 Things that are surprising, strange or incongruous may foster noticing.

3.11.1 This summer when I was in Minnesota camping, we spent all morning one day going up a creek. Suddenly, around midday, we came upon this huge machine right in the middle of the stream. We had no idea what it was. We'd gone though a lot of interesting territory but hadn't noticed much. When we got to this machine, however, it stopped us. It seemed out of place and took our attention.

3.11.2 I notice things that are different than usual – like rain. Most of the time I don't notice what the weather is like, but on a rainy day you can't help but notice.

3.11.3 I noticed the water level of the pond in the park. Usually it's not so low.

3.11.4 I notice when I cross into Delaware – the road's surface and colour is quite different from that of New Jersey's.

3.11.5 Yesterday I was driving and I looked in my rear-view mirror. I looked again. The car behind me was without a hood!

3.12 Things of beauty or unattractiveness may evoke noticing.

3.12.1 I was walking to class. All of a sudden my eye caught a sparkle of light. I looked up and was surprised to see how pretty the light from the snow-covered trees was.

3.12.2 On the way to Washington, D.C., everyone in the car noticed a field of pumpkins. There were rows and rows of them in the field. They looked beautiful, resting there in the field.

3.12.3 I always notice the bank building on the corner when I'm walking. It's round and hideously designed and it catches my eye because it's so different.

3.13 Things in which people have a personal stake or interest may evoke noticing.

3.13.1 I was in a class last year that studied utilities distributed through wires. Now I notice quite often if telephone and electric wires are overhead in a place, the types of poles used, their age, the aesthetics of construction – things like that.

3.13.2 I never used to notice coloured shadows – in fact, I never knew they existed. Yet because of a course I took that spent time studying them, I've become aware of coloured shadows and look for them when I think of them. I notice them often now, especially in the streets at night. The best thing is the more I notice coloured shadows, the more I look for them. At first, I was aware of them only rarely, but now I notice them quite often. It's not that I walk down the street saying to myself, 'Okay, it's time for you to be conscious of coloured shadows.' Rather, the thought of them will suddenly 'pop into mind', something in me will look, and maybe I spot one. Or sometimes I'll be walking along 'in a daze' and suddenly I notice one. They jump out at me – I don't make any active effort to see them. It's as if they show themselves to me and I don't do a thing but respond to them.

3.14 Sometimes people don't notice the world until it has become something different.

3.14.1 I was walking to the store to get milk on Saturday. The street seemed different. Suddenly I noticed the reason: someone had cut down a large tree. I hadn't really noticed the tree while it was there, but now that it was gone, the scene felt different.

3.14.2 The other day I walked into a professor's office which I hadn't been in for a while. Something about it felt different but I couldn't tell what at first. I suddenly noticed that the difference was due to the blackboard: it was clean for a change! Usually it's chalky black.

3.15 Mood and energy level affect noticing.

3.15.1 I had a good talk with a professor and I was feeling happy because of it. I took a walk over by the pond. It seemed I was noticing a lot because I was feeling so good. The ducks on the pond, the colours reflected in the water, the trees – I was very much aware of them. I was noticing more around me than I usually do.

3.15.2 I developed some negatives I had taken last week in my photography class this morning. They came out well and I was encouraged. I felt happy and it gave me so much energy that I spent the entire afternoon taking more pictures. I had a good time; it was one of the best picture-taking sessions I've ever had. I felt a part of the things I was taking pictures of. I was noticing more than I usually do, and it had something to do with the fact that the photographs had come out so well.

3.15.3 I took a bus home to Virginia and was let off about three-quarters of a mile from my home. This is a walk I'm familiar with and

I had been looking forward to it on the bus, excited about old sights and memories. When I got off the bus, however, I noticed how miserable I felt, tired and hungry. I just wanted to stop travelling. About half the walk was over before I noticed that everything was passing me by. I wasn't making connection at all. But five seconds later I had 'drifted off' again. I was so 'tuned out' that I went trudging across some grass that was a short-cut. I didn't notice it until the deed was done and then had a good laugh at myself.

3.15.4 I know that when I'm in a bad mood I notice smells more, especially if they're unpleasant.

3.15.5 When I'm upset I notice people's bad qualities more. I become aware of qualities in people that wouldn't bother me otherwise.

3.15.6 Last Thursday afternoon I went grocery shopping. Everything seemed wrong. The store was out of some things I needed, the cashier wouldn't accept my foodstamps because I didn't have my identification card. I was angry. Everywhere I looked I saw another thing which showed in more detail what a mess the world was in. I remember noticing the four big metal pillars that hold up a huge electric sign for the supermarket where I had been shopping. There was this quick flash of annoyance in me that was saying, 'What a waste of resources! – all that people in this crazy country know how to do is waste.' I was startled to see myself upset by so little a thing as a sign, but I couldn't shake my negativity. I went home and went to bed.

3.16 At special moments people may have an intensity of encounter with the world outside themselves.

3.16.1 I was sitting on my usual bench, facing the two brown houses and pine trees behind the court where I usually play tennis. As I was gazing over at the pine trees, the sun went behind the clouds for a moment and some noisy birds flew over the houses. Suddenly, I felt very still but shivery inside. I felt quiet, I felt as if all the world was at peace. I felt warm toward everything.

3.16.2 One day this past summer I was driving across the Verrazano Bridge. All of a sudden I felt very high emotionally and in harmony with everything around me. The bridge stood out as a strong, all-consuming structure, yet at the same time I felt connected to the bridge in some kind of spiritual way. The moment lasted as long as I was on the bridge. It was vivid and I clearly remember it.

3.16.3 On Wednesday I visited a museum to do some research on the Shakers. I had driven quite far on the forested back roads, but the whole trip I hadn't really noticed anything. I was caught up in my thoughts. When I reached the museum I drove the car down the long driveway and parked it in the lot which overlooks a valley. As I got out, I had this strong experience. I felt a rush of warm spring air on my face, I breathed it, and then stood for several seconds overlooking the valley before me. I suddenly understood who the Shakers were and why they had chosen to live as they did. For the first time during my months of research, I felt that I could understand their love of order and beauty. It was as if I felt the heritage of the place pass through me.

3.17 A relationship exists between at-homeness and encounter.

3.17.1 When I came to Worcester, I moved into an apartment in which I wasn't comfortable because of my room-mates and the 'dumpiness' of the place. I definitely noticed that in the time I was living in this unpleasant situation there was no energy to spend on new activities. I was upset by the living situation and had no interest in doing anything other than the basic necessities. One of the reasons I had returned to college was to grow as a person – to try out new things. But the apartment situation upset me. I had no wish to get involved in anything. I did just the minimal to keep me going. After I changed apartments and began living with people I liked, I felt more comfortable again. There was more energy. I could give myself to new things and take an active interest in life again. For example, I got involved in a pottery course, and volunteered my help at a nursing home. I felt more free inside and could get involved with things outside myself.

3.17.2 For five years I lived in the same city. I was very much at home there. I had everything so easy. I had a job, friends, a nice place to live. The problem was that everything was *too* nice – life seemed stale. The same schedule day after day, the same people – everything was usually the same. I saw how I could easily live like this the rest of my life. I thought, 'You've got to get yourself out of this rut.' I decided to make a change by returning to school.

APPENDIX B: COMMENTARIES ON THE CLARK ENVIRONMENTAL EXPERIENCE GROUPS

The following commentaries discuss the value of the environmental experience groups as a learning process. Those which were given to me in written form are marked with an asterisk. Although I asked all participants to either speak with me about the group experience or describe its impact in writing, not all did. I include here all commentaries I received. Editorial changes have been made to improve flow of text.

Commentary 1*

I'm excited when I think of what last semester's environmental experience group meetings have done for me. I don't know if I'll be able to express all that I've gotten from the group, but I'll try. I'm gaining a deeper, fuller understanding of myself through an awareness of the factor that seems to have a great influence over me – my experiential relationship with my environment. My mind and body are slowly becoming attuned to aspects of movement and rest that I hadn't noticed before, and to the significance of space. I am more fully aware of certain wants and needs now than I think I was before. Unconscious realms are becoming more conscious ones: it excites me more and more as I realise how much I take for granted in my everyday life.

There's a beauty in my body's activities that I've never seen before, manifested in flowing order and choreography. I sometimes feel like a whole new world is opening up for me in terms of places and situations – my home, my school, my need for other people and my need for privacy. I appreciate all these things with a deeper understanding of their necessary role in my life. I want to share my excitement. The thought of helping other people understand these things – especially children – brings purposeful meaning to me. I feel like I've never felt or seen these things before.

Commentary 2*

Our discussion group made me more aware of many of my movements and habits which I had previously dealt with through 'body knowledge', but always implicitly. It was both stimulating and unnerving to discover how many assumptions we unknowingly make about our interactions with our surroundings.

I had never before thought about the ways in which surroundings not only alter our emotional state, but also shape our behaviours. Previously, I had known that emotional states are reflected in the way we organise our living spaces, but I had not been aware of the concomitant effect of that organisation on emotional states.

The group was also a valuable didactic setting; I learned about the general characteristics of a phenomenological approach and I began to understand some of the concerns of geographers.

Commentary 3*

One of the most important things about the work we've done for me is that it has shown me a way of knowing that I can respect. In catching phenomena, not manipulating them, and sharing these observations with each other, I feel as if we've been receptive and open to the thing itself. So many of my friends feel a gap between what they're exposed to academically and the rest of their (more relevant) learning experiences. For me, the work we've done in the group resolves this problem. I don't feel a split; learning about myself has been a part of my education here at Clark because of the groups. That is the strongest force which keeps me coming back to school.

Glimpsing completely unexpected, very basic forces that shape my behaviour is exciting. Seeing a little more fully that I'm not at all in control of many of my everyday movements, feeling that this is important to my own experience in a lot of different ways – like understanding a hard-to-break habit such as nail-chewing, or working on a new piano piece and being able to tackle difficult passages more effectively, or in adjusting to a new living situation like this past summer – being able to recognise the importance of routine, of centre, of familiar paths, and being able to encourage their development.[1] For instance, one thing I noticed over the summer was the way the house I was living in became a home. There was a sense of centre there that survived moodiness and restlessness. How important that was! All sorts of little things helped – someone lent me sheets, so I didn't have to sleep in a sleeping bag any more, and I moved the bed back to the bedroom – it was in the living room when I came – so that there was a living space and a sleeping space. Having this home, as temporary as it was, ended up to be one of the best things about the summer. Real delight in cleaning things up, putting things in their places, having a place that was mine.

What was especially interesting to me this summer was the

phenomenon of obliviousness. Catching myself (mostly afterwards) so many times completely caught up with some aspect of the situation, so that I forgot about the outside world and felt afterwards as if someone else had had the experience. I observed that sometimes the obliviousness had to do with a goal I'd set. For example, one day I took a hike up Pelican Canyon, with ambitious plans to take another trail down a different canyon on the way back. With my day all planned out, the actual hike became almost automatic – an empty gesture. I caught myself at one point completely preoccupied with random thoughts. The whole tenor of walking the trail changed for a few minutes, as if I'd only just arrived, aware of surroundings that I'd been blind to moments before.

It was a frustrating day, with most of it spent rushing to the next landmark. I see this obliviousness orientation in so much I do and often wonder if there's some way to keep myself more in touch with the moments at hand. Here is where the group work has been helpful because in the past I wasn't even aware of this obliviousness – at least now I see it and perhaps in time I can find more ways of getting beyond it, of actually seeing and looking at the things that are there with me at the moment.

Commentary 4*

I feel that the group has been, in many ways, one of the most important experiences of my academic career in so far as it has given me new insight and awareness into myself and my dealings with the everyday world. Self-knowledge is a kind of learning I value highly. I think that our probing into our everyday experiences and the feedback from other members of the group brought out clear patterns of movement, centring and encounter which I took for granted before. It is not that our discoveries are so esoteric that they can't be grasped by others: it is the reverse. We have found patterns that are comprehensible because of their universal nature. That is, we can (perhaps) put forth some general assumptions as to man's role in his day-to-day world: for example, we are centring beings, and much of our movement is habitual and semi-automatic.

Commentary 5*

I feel that since the beginning of my participation in the environmental experience groups, I have become more aware of two things – noticing my own experience of my surroundings, and generally trying to relate to other people patterns of environmental experience and the phenomenological method discussed and discovered by our groups.

I always like to apply things I learn in school to my everyday living. When I learn that people do things by routine a large proportion of the time, or that there exists a theory of 'body knowledge', I try to notice the times I or someone else is operating routinely or I find that I can draw upon body knowledge as a possible explanation of a specific behaviour. I do not consciously make an effort to apply these concepts; they just occur to me when I try to understand something or just observe people's behaviour.

I guess that what I am trying to say is that the group work has carried over and become a part of my every day experience. I think this started due to the weekly noticing tasks. All during the week I'd be aware of looking for my behaviour, regarding the specific topic, but I'd also notice the week's themes previously and just generally everything we've discussed. I began to notice other people's behaviour, too, and in this way, I find that I've integrated body knowledge and concepts of noticing (or not noticing), centring, routine ballets, etc. into my stockroom of approaches to understanding, describing and explaining behaviour. I just think that way now. I notice without trying or planning to notice behaviour.

Because I've become interested in applying these new (for me) ideas, I enjoy going to the group meetings and sharing experiences, reacting to and getting reactions from people who I feel understand and share my interest and enthusiasm. I do feel a certain special tie to the people in my group – as though we're discovering something personally and all together, at the same time and in the same way. I feel there is a special rapport, probably because we share *experience* rather than intellectual knowledge or ability, like most school group meetings and classes. This is why I have continued: I value both my interest in the subject as well as my affection for the people involved, who I probably would never have met in any other way.

Commentary 6*

One of the positive features of the environmental experience group is the nice feeling of meeting with the same people every week and discovering that others sometimes act the way I do in my own situation. It's interesting to discuss things we do in our everyday lives without usually thinking about them. The discussions have helped me become more aware of how structured I make my own life and how the 'ordering' of things and people can be a bodily and emotional process.

For example, right after vacation I moved from one dorm to another. When putting things away, I realised that I was organising books, clothes, etc., in a way similar to the arrangement of my old room. This led to the affirmation of the 'logic' of the placement of things in my first room.

Commentary 7*

About a minute ago I walked into my room at home, thought over the question of writing this report or not, and decided not to. The next thing that crossed my mind was 'What time is it?' I looked over to where the clock was last night, and to my surprise, it was not there. I immediately remembered that I had moved it closer to my bed so that I could hear it better. I laughed hard, as I realised that this was a good reason to write up my observations concerning environmental experience. I think the example of the clock, if stretched a bit, portrays accurately the newly gained level of knowledge that now guides my view of environmental experience.

I was new in the Clark community when I began the group experience, so I was able to notice how I set myself up in the environment as it was taking place. The observations I made of my own behaviour were no different than the others in the group, but I was seeing my interactions develop, as opposed to reliving everyday dealings with the environment (which mine were soon to become). I'm sure the minute I arrived at Clark I began to order things. I established walking routes to buildings, and chose a particular side of my room and then arranged it in a particular way in terms of which I could eventually move habitually – without having to think where things were.

What all this points to, I think, is that through certain exercises done with the group, I developed a certain perspective regarding myself and my environment. Knowing now, through observation and practice, that we take so much for granted about our environment, I can be conscious of these dealings and so participate more actively with my environment. I can take control of it more easily. That is not to say that I can avoid completely subjective development of a home, centres, habitual paths, etc., but that I can be aware of the possibilities of walking different ways to places and not immediately and automatically walking the same way each time, or I can begin to see consciously how attached I have become to a particular place. Changing the position of the clock was an example of taking part in the construction of an environment that I was in at the time. Once a person attunes himself to his dealings with his environment he will be less likely to resume his old ways of dealing with it. He has more of a choice in the way he conducts his day-to-day dealings because he is now aware of them more than he was in the past.

Commentary 8*

Unfortunately, I wasn't able to attend the meetings regularly for the second half of the semester, and am slightly at a loss as to what went on. I was very intrigued by what I did get into; it made me more conscious of my movements in general. Especially good was the week we spent on home as a centre; it helped me to crystallise my own thinking on my need to be centred and my feelings while on the road. It also brought into perspective the last two years during which I lived in four cities and even more apartments, dorms and houses, and spent a good deal of time travelling.

Commentary 9*

I joined the environmental experience group not really knowing what it was all about. Right now I am further along, but I'm still not exactly clear about what's going on.

The outside exercises (rearrange your furniture, go a different way, etc.) were one of the most enlightening parts of the group. These combined with the weekly discussions gave me a greater awareness of my day-to-day life. Whether this is good or useful I don't know. I do know that it can be a pain to realise the significance of everything that I do, but sometimes this gain in knowledge helps me to re-examine and evaluate my actions and adjust them accordingly.

Commentary 10

The group has helped me to fit things together – for instance, territorial space and rearranging things. It's given me a chance to explore things, to clarify things a bit.

Some of the themes were more helpful than others. The task of walking a different way than usual helped me to think about where I go in a day and how I always use some routes and don't go other ways. I started noticing what I might be missing. The idea of centring is something I've been thinking about a lot lately because I'm going to move off-campus. I'm wondering where my centre will be – the apartment won't be like the dorm, where there are always a lot of people. Will I hang around other places as opposed to my apartment? The group has really made me think more about things I had thought about and think of why I do them the way I do.

Commentary 11

I've definitely gotten a lot out of the group. I think before we started I was looking at how I lived similarly, but here I had a directed way to do it and things made more sense. Last night I was reading something I had written on territoriality at all levels. I had a simulated model, but because of things I've learned in the group, I would never look at territoriality like that any more. The model was a creative idea, but now I don't think it proves anything. I think there's something in territoriality, but not in the way I did it. I thought that personal scale and national scale were all the same, and now I think there are many differences. I know a lot more about personal scale now. I don't see how it could work on a national scale. Some of the needs are the same if you blow it up enough, but I don't think the comparison gets you too far – as I had thought before. The group has helped me to see that there are different things going on in the two. Most people in their everyday lives have little consideration for the country they're in as a political unit – unless they're forced to consider it – if you're going to be killed or if you're of the 'wrong' political party. But generally, people are just concerned with their everyday lives. I don't think they think in terms of political boundaries – in terms of 'Because I'm an American, I'm more secure.' They think they're secure for other reasons. I don't think people think of it too much on a regional basis either. Of course, if I'm living in the East that may affect the way I think, rather than if I'd lived in the Midwest, but that doesn't really have much to do with the way I live everyday.

The groups helped me to see this because I was looking at myself in a more organised fashion. I always thought that how I lived my life didn't have much to do with politics and national events. The group gave me a clearer outlook – a clearer way of thinking about such issues. I can't say before coming in the group I had never looked at the things we looked at, but I don't think I had looked at them as clearly.

One of the most significant themes we looked at to me was order – it seems to have come into everything. For example, in terms of having a place for everything, one of the things I want to do when I go home is to organise all my papers, books, and find out where everything is. Centring was helpful too. Noticing things like, 'I'm going home now – ah, yes, the centre.' I noticed it in terms of visiting, too. I visited a friend in Ottawa, and the first thing I did was to go to her bedroom. That was my centre – everything seemed organised around that bedroom. Once I was there I had everything. I was conscious of that feeling while I was there. Centring can have an effect in terms of knowing what you're doing and being in touch with yourself. When I've gone to New York to visit my friend, I've realised how much I've hated it – and it's because my friend knows the city well and I don't. When I'm with him, I don't know where I am, I'm totally disoriented. Yet there are times when I go to New York when I like it, because then I have some sense of organisation of the place. I think it's because then I have to get around by myself. It was the same in Ottawa – it was more exciting because I had a car and I was doing the driving so I could get to know the city. And I felt better. There's a control you have.

The themes got more difficult – they got to be the kind you could think about. They didn't work well for a group because they weren't concrete examples any more. It's not like 'I'm going here and there' – without thinking about your experience you can't say what it is – it's more philosophical. I think they were authentic themes. Especially centre-horizon. For several days I hadn't been home much at all: other places, other people's houses. And then suddenly there was this need: I have to go home. You want to be active, and then you want nothing.

I think it's important to know how you live. What we've been doing in the group is looking at the part we normally take for granted and in that sense it's helpful. I think I am more aware of what I'm doing, of where I go, of why I am here, of obliviousness. I was aware of these things before, but I'm more aware of them now. I'm aware of body-subject, too – in terms of forming habits. Like in terms of organising my papers – I'll have all my papers and books, everything I'll need, in their places, and the places will become habitual. Because I'm now consciously aware of the force of habit, I think I'll be more organised in the next place I live.

I'd like to do more of this work in graduate school. I think I can, but I'm not exactly sure. I know with planning, controlling pollution – I know they consider all the wrong things. And I think what we've discovered can be used. I still think what we're doing is theoretical. I think it's very valid, but in the end we're doing the same thing as everyone else, based on different dimensions, but still a theory. I think it's better – this one is real – it occurs, this is what's going on. You can prove anything – it's like giving the monkey marijuana for four days and then concluding it's bad. Whereas here, we'd have to take people as they regularly smoke it, how it affects them in a day-to-day context. It's a model, but it's more in contact with what's going on than many other social science theories that I've learned about.

Commentary 12

The group was useful – I'm thinking about things differently all the time. For example, take my job in the hospital. I noticed a lot of things about how people there got around that I wouldn't have noticed before the group. The people in the hospital weren't willing to learn more than one daily route pattern – they just want to get out and in. There's a beautiful library with comfortable furniture in the building, but most of them never have been in it because they have a regular routine and don't like to vary it.

I seem to be on the look-out for the same kind of experience we talked about in the group. I'm trying to understand body-subject – mostly I notice things about buildings rather than outdoor things. It took me a few weeks to get the hang of habit and the possibility of bodily knowledge – it was hard to notice without interpreting it immediately. And it was hard to be open to observing – I'd never done it before and at first I wasn't open to it. It just wasn't a way of learning that I had experienced before.

Centring I understand and I feel it strongly. I was aware of it in the past, but I didn't call it centring. I made my room a centre so that I'd have a place to go where people wouldn't come unless I wanted them. When I went home, I didn't have much of a centre. Like the living room – no one sits in there because it's my mother's and it's for guests, and you don't want to dirty it. But when I was home last I spent a lot of time in there reading – it became a kind of centre – though my mother would move my books.

The last themes weren't as helpful – they were vague, and not interesting.

Commentary 13

I didn't get much out of the group. A little. It was funny – the first couple of sessions I really 'got into' it – doing the maps of our daily movements, looking at things more closely. My movements fascinated me – to see how they shaped up on the map. As time went on, I didn't do the themes as a regular thing. I tried to integrate the themes with other things but it was hard – fitting them in with my work. I tried to keep observing in my day, but I was doing it less and less. At the end, I was pulling experiences out of memory. Some of the themes were more difficult than others. I couldn't cope with the one 'deciding where to go when'. There was noticing too. I didn't get a feel for it either.

I didn't want to analyse my behaviour – I just gave, hoping what I said would be helpful. I like to talk. I don't know if I'd do the group again. I was somewhat passive in it. I like being in psych experiments, I like being a subject. This required me to be more of an experimenter – I found I couldn't think clearly about many of the themes. I like things to be more tangible and graphic. This is more philosophical.

Note

1. This commentary was written in the autumn of 1975, after the commentator had gone to Utah for the summer to work as a Forest Service guide at Fishlake National Forest.

APPENDIX C: ORGANISING AN ENVIRONMENTAL EXPERIENCE GROUP

This appendix provides a set of step-by-step instructions for the student who wishes to investigate more closely his *own* everyday geographical experience, which in ordinary day-to-day living he takes for granted. This appendix will be better used if its procedures are actually *done*, rather than just read. Even if the student involves himself in this project for only a few weeks, he may gain important new perspectives on his everyday experiential relations with space, place and environment.

Step One: Organising an Environmental Experience Group

Although the interested student could investigate his own everyday experience alone, group exploration is much more effective because it generates a collection of observations the number and variety of which no single person could muster through his own efforts. In terms of size, the experience of the Clark group indicates that a group of between five to eight participants is good. A smaller number may reduce the spontaneous flow of observations, while a larger group may restrict the chance of participation by all members.

The Clark group met one and a half hours weekly in the evening, and this schedule worked out well. As long as there was a small core of people who attended regularly, the meetings were not overtly disturbed by other individuals who came irregularly. Generally, however, the people with steady attendance benefited considerably more from the group experience than those who came only sporadically.

To provide the group process with direction, one person should assume responsibility for guiding the meeting and asking relevant questions at times when there is an unclear observation, lull, or other awkwardness. Following each theme (see below) are a set of questions which may aid discussion and interpretation of the observations.

Step Two: Making Use of the Themes Used in the Clark Environmental Experience Group

A pure phenomenology imposes no bounds on experience: all experiential reports on all human situations would be of equal importance and reveal in greater and greater detail the underlying dynamics of life's taken-for-grantedness. As geographers, we limit our interests to the world of geographical experience, but even this incorporates such a vast nexus of experiential parts that we are required to find some method for organising this complexity into a simpler framework which still maintains authentic contact with the original world of concrete experience.

In the Clark group, this ordering was sought through the use of a theme, which organised each week's meeting around a specific topic that earlier meetings had indicated might be significant. Individuals creating a new group might try these themes and supplement them with other topics that seemed important.

The seventeen themes used by the Clark group over a period of two semesters are articulated below. Group members should seek to discover concrete situations in their own daily experience which may have relevance to the particular week's theme. They can then report their experiences at the weekly theme meeting so that other group members can share in their discoveries and verify their accuracy.

Step Three: Interpreting the Observations

At least for the first few weeks, group members should simply report observations without attempting interpretation. Ideally, observations should be transcribed weekly and placed on individual cards which can be reviewed by members. This possibility, however, requires considerable time and expense. Alternately, one member of the group should take brief notes on the observations in theme meetings. Then after the group has met for a few weeks (four to five weeks is probably a good number), it can devote one meeting to a review and discussion of themes and observations. At this time, too, group members might wish to compare and contrast their discoveries with observations provided in Appendix A of this book.

As the group proceeds, members will more than likely notice connections and patterns which were unnoticed before. On one hand, the discoveries provided will not be new – only specific illustrations of similar events which one has experienced time and time again in the past. On the other hand, this group procedure seeks to penetrate these well known, mundane situations and discover underlying similarities and connections. Out of these new discoveries can arise an expanded awareness of what it means to live in a geographical world.

Theme One: Everyday Movement Patterns

For the coming week, take note of your everyday movement routines.

Observe the various places you visit each day in the order which you go to them and when you can, make a list of this pattern at the end of the day. Ideally, group members should perform this task every day during the week, but even a single attempt for just one day will provide important information and observations.

Key questions to keep in mind at the theme meeting:

1. Are there any regularities in your day-to-day movement patterns in terms of time and space?
2. Are there any crucial places and routes in the geographical space through which you move every day?
3. Are there any kinds of experiential space-time dynamics in your everyday movement patterns?
4. How could your daily movement pattern be represented most simply?

Theme Two: Centring

In this theme we wish to consider if our lives are centred spatially – do you have a centre or centres in space? If so, what functions do these centres provide? Why are centres needed? Is quality of life weakened for a person who doesn't have a centre? Have you know different centres in your life? How do varying life-styles affect centring? Is there more than one centre in your everyday situation?

Key questions:

1. Do group members report the presence of centres?
2. What needs does a centre serve?
3. What kinds of centres are there?
4. Is centring a crucial part of everyday human experience, or can it be sacrificed without harm to the person?

Theme Three: Noticing

This week we wish to observe the situation of noticing. When do we notice things outside of ourselves, especially things in the physical environment? When do we make contact with what we see, when do we hear a sound, when do we notice a smell, etc.? What can be said about the state of noticing? Are we noticing all the time? If not, what's happening to us when we're not noticing? Why do we notice something at one time but not at other times? Are some people more aware of the world outside than other people? What is noticing dependent on? Concrete observation on any of these questions will be extremely revealing.

Key questions:

1. What triggers noticing? What features in the physical environment? What conditions in the person?
2. Are there intensities of noticing? Can one's contact with the world outside vary, depending on the particular moment?
3. Is there an experience in the human situation which is the exact opposite of noticing?

Theme Four: Moving in Space

How do we get about? What is the process like? Do we take the same routes to the same places all the time? What routes do we use to get where? In short, what is movement as an experience about?

Do-it-yourself experiment: Set yourself for one day (or more) to go to a place by a different path than you usually would. To assure more than one experimental attempt, choose a place that you go to several times in a day. Can the experiment be done? What is the experience like as you go the other way?

Key questions:

1. What part of ourselves is responsible for everyday movement?
2. What happens when routine movement patterns are upset?
3. Is everyday movement habitual in any way?

Theme Five: What Do We Pay Attention to as We Move through Space?

The preceding theme considered how we move through space. In this present theme, we seek to clarify in more detail what is happening as we move. For example, what do we notice as we move, what do we fail to notice? How do different transportation modes affect our movement experience? What effect does route type (sidewalk, back road, rocky path, expressway) have on our movement experience? Clearly this theme is closely related to both noticing (theme three) and movement (theme four), yet is different enough to be considered separately.

Key questions:

1. What things affect our noticing as we move?
2. What do we pay attention to as we move?
3. Can the movement experience be broken down into components?

Theme Six: Emotions Relating to Place, Space and Environment

Do we attach particular feelings to particular places? At particular times do certain environments or things in those environments evoke an emotional response in us? Are anger and annoyance emotions that can relate to place? How about happiness, contentment, peacefulness? Are there feelings that you associate with places where you spend a lot of time? Does the old cliché 'Home is where the heart is' have any experiential significance? If you can, try to catch feelings as they happen.

Key questions:

1. What range of emotions in regard to place was registered by group members?
2. What are the common characteristics of places which evoke emotional response?
3. How do emotions relating to place arise? Does it take time for a person to become emotionally connected with a particular place?

Theme Seven: A Place for Everything, Everything in Its Place

Yes, it's a cliché, but we wish to find out if it has any experiential significance. What is established for us if everything has its place? Does the body perform better when places are provided for things? What is the relation between habitual routine and place? How important is it in your life to have places for things? What things do you put in what places? At what environmental scales is this theme important?

Do-it-yourself experiment: intentionally move one thing that has a permanent place to another place. For example, change the location of a towel, an alarm clock, a favourite chair – whatever. Observe your reactions.

Key questions:

1. Do you discover an ordering pattern at work in terms of places for things?
2. At what environmental scales do you find 'place for things' important?
3. Does an ordered environment enhance the quality of everyday living in any way? Does it detract in any way?
4. Do you find a kind of mutually sustaining dialectic between routine behaviour and places for things?

Theme Eight: Deciding Where to Go When

Past themes have focused on two dimensions of spatial movement: the process of moving itself and the experience of what we notice as we move along. Now we need to probe the nature of the decision to go. What happens at the moment when we decide to go somewhere? Is this decision always of the same sort? To facilitate this theme, it may be helpful to focus on a few specific activities, such as going for a drink of water, going to the bathroom, going for food, going to class, going to a place you never usually go to, etc. The most useful observations will be those which catch the nature of the decision as it happens.

Key questions:

1. How does the decision to go happen?
2. Is the decision to go always of the same sort?
3. How is the decision related to actual movement and to the destination of the trip?

Theme Nine: Off-Centring

Our discoveries so far indicate that through repetition the body often establishes set patterns, for example, in terms of movement and places for things. This week we wish to consider the role of position in terms of the body itself. In other words, do we establish specific positionings and locations for ourselves? For example, when we position ourselves in front of or behind something, do we always locate ourselves in the same way in the same place? Do we become accustomed to a particular positional and directional orientation in variously scaled environments that we use?

Do-it-yourself experiment: consider your normal positionings in terms of sinks, shelves, tables, rooms, outdoor spaces. Then, seek to change one of the positionings. What happens, how do you feel, how consistently can you do it?

Key questions:

1. Did you notice the importance of your own habitual positionings in your everyday movement? At what scales?
2. Can you think of times when a change in body form – for instance, a new pair of glasses or a knapsack – has disturbed the way in which you routinely position yourself?
3. What happened when you did the experiment? Were you able to

do it? Did the task create any tension in situations that you usually take for granted?

Theme Ten: Destinations

Themes on everyday movement have focused so far on how we move in space, what we notice as we move, and how we decide to go somewhere when. Another important aspect of movement may be the role of destination in everyday travels. Is a destination important? How does a trip without a destination differ from one with a destination? Can there in fact be a trip without a destination? Is there a difference between purpose and destination in everyday movement?

Key questions:

1. Did you notice any trips without a destination?
2. Is the notion of intentionality relevant to your observations? Did you notice any trips that were purposeless?
3. Are there other scales of movement at which purpose is important? For instance, on a smaller scale – in movement around your home, or around a room?

Theme Eleven: Disorientation, Getting Lost, Being Lost

We've probed the importance of order in our everyday experience – in the paths we use, in the places we put things, in the way we position our bodies in different situations. For the next week, it may be helpful to consider situations in which this order is upset, or hasn't yet been established – when we're lost or disoriented. In addition to trying to catch the experience as it happens, you may want to think back to times in the past when you've been lost, and to talk to other people about their experiences. Have you ever really been lost? In what contexts and environments do people get lost? Can one be disoriented without being lost?

Key questions:

1. In what different situations has the order in your everyday movements been disrupted?
2. Does the design of the physical environment play a role in ease of orientation? For instance, can you think of any buildings or places that are difficult to navigate?
3. Are there certain ways in which you try to regain your bearings when you are lost? Disoriented?
4. Is it possible to be disoriented in time as well as space?

Theme Twelve: Care and Ownness

Another taken-for-granted aspect of our relationship to place may be an attachment to place. What places do you care for? How does this care arise? Are there places that belong to you in some kind of personal way? How do you make a place your own? Are there places that exude a spirit of care, or places that no one cares for?

To facilitate thoughts on this theme, you may want to make a list of the places in your life that have meant a lot to you. You might also speak to other people about places that they care for.

Key questions:

1. Are there types of places that appear on everyone's list – for instance, homes?
2. Is there a relationship between experience with a place and care for it? How about noticing?

Theme Thirteen: Obliviousness and Immersion

We probed the notion of noticing in the third theme. For next week, let's try to look at what state we are in when we are not actually noticing things in the world outside us. What are we doing when we are in this state? Are some people more often oblivious than others? What role does destination in movement play in obliviousness?

Key questions:

1. Were you able to catch yourself at any point during the week when you weren't making contact with your surroundings?
2. Is there a connection between routine movement and obliviousness? That is, does familiarity with a task or route encourage us to do it automatically?
3. Is obliviousness a liberating and/or constraining force in your life?

Theme Fourteen: Paths, Attachment to and Points Along

In themes for past weeks, we've looked at the nature of our movements over paths and at our attachment to certain points in space. Let's focus for the next week on attachment to place in relationship to our habitual paths. Do we become attached to paths? Are there any routes that you've known that have a special significance for you? Are there any that you have bad feelings for? Are there places along paths that are meaningful for you? One example of this might be indicators of approach to destination.

Key questions:

1. Did you discover that some paths are more comfortable than others? What factors help to account for this?
2. Does a path's length seem the same to you in both directions? Do you have different indicators of approach to destination depending on your direction?
3. What sorts of spots along paths have special significance for you?

Theme Fifteen: Order

Throughout our meetings, order has been a strong and underlying theme that has run through many of our observations. For next week, we need to look in more detail at just how order manifests itself in our everyday lives, and how we manifest order in our everyday lives. What is order? How does it extend itself? How do we create it, and what happens when it is not present?

To facilitate this theme, let's all try to consider in what ways – especially as we relate to the geographical world – order pervades our lives. Think of this in terms of scale – bed, room, building, outside spaces, etc. You might also think of it in terms of time and social relations, and your inner situation.

This is a difficult theme that can easily move into abstractions, but hopefully we have a solid enough concrete base to come up with some genuine contact with order's significance for people.

Key questions:

1. At what different scales is order important in your everyday experience?
2. How does order act as a constraining or liberating force in your life?
3. How is order created? How does it maintain itself?

Theme Sixteen: Spring*

We haven't looked very closely yet at the natural environment, effect of weather, and other environmental phenomena. For next week, let's take 'spring', and see if any sorts of things come from it. How is spring noticeable? What does it mean in terms of what we are?

* There is no discussion of this theme in the interpretive part of this book. The theme reflects an aspect of environmental experience – seasonality – which demands its own detailed phenomenology.

Key questions:

1. What things are signs of spring for you?
2. How does the change in weather affect your experience in space?
3. Do you find that any of your routes have changed with the change in weather?

Theme Seventeen: The Tension between Centre and Horizon

The architect van Eych suggests that 'man is both centre bound and horizon bound'. For the coming week, let's consider this theme, which has been suggested by many of our other topics. Do we fluctuate between centre and horizon, between security and strangeness, between peace and confusion? This is a difficult theme, and will probably involve more thought than observation. Our portrait of geographical experience needs this overriding framework, which seems to cloak the other patterns that we have discovered like an invisible umbrella.

Key questions:

1. What other polarities are at work in your experience?
2. Does order play a role in the tension between centre and horizon?
3. Do you find this tension manifesting itself in different ways in your life?
4. Do you notice any patterns (seasonal, weekly, daily, etc.) that reflect this interaction?

REFERENCES

Adams, John S. 1969, Directional Bias in Intra-urban Migration. *Economic Geography*, 45, pp.320-3

Allard, A. 1972, *The Human Imperative.* New York: Columbia University Press

Allport, F.H. 1955, *Theories of Perception and the Concept of Structure.* New York: Wiley

Appleyard, Donald. 1970, Styles and Methods of Structuring a City. *Environment and Behavior*, 2, pp.100-18

Ardrey, Robert. 1966, *The Territorial Imperative.* New York: Atheneum

Bachelard, Gaston. 1958, *The Poetics of Space*, trans. Maria Jolas. Boston: Beacon Press

Backster, Cleve. 1968, Evidence of a Primary Perception in Plant Life. *International Journal of Parapsychology*, 10, pp.329-48

Bannan, John F. 1967, *The Philosophy of Merleau-Ponty.* New York: Harcourt, Brace and World

Banse, Ewald. 1969, Historical Development and Task of Geography. In *A Question of Place: The Development of Geographical Thought*, eds. Eric Fisher *et al.*, pp.168-74. Arlington, Virginia: Beatty

Barbour, Ian G. (ed.). 1973, *Western Man and Environmental Ethics.* Reading, Massachusetts: Addison-Wesley

Barral, Mary Rose. 1965. *Merleau-Ponty: The Role of Body-subject in Interpersonal Relations.* Pittsburgh: Duquesne University Press.

Beck, Robert J., and Wood, Denis. 1976a, Cognitive Transformation of Information from Urban Geographic Fields to Mental Maps. *Environment and Behavior*, 8, pp.199-238

——, 1976b, Comparative Developmental Analysis of Individual and Aggregated Cognitive Maps of London. In Moore and Golledge (eds.). (1976), pp.173-84

Bennett, J.G. 1966, *The Dramatic Universe, vol.3. Man and His Nature.* London: Hodder and Stoughton

Boal, Frederick J. 1969, Territoriality in the Shankill-Falls Divide, Belfast. *Irish Geography*, 6, pp.30-50

——, 1971, Territoriality and Class: a Study of Two Residential Areas in Belfast. *Irish Geography*, 8, pp.229-48

Bollnow, Otto. 1967, Lived-space. In *Readings in Existential*

Phenomenology, eds. N. Lawrence and D. O'Connor, pp.178-86. Englewood Cliffs, New Jersey: Prentice-Hall

Boorstin, Daniel. 1973, *The Americans: The Democratic Experience.* New York: Random House

Bortoft, Henri. 1971, The Whole: Counterfeit and Authentic. *Systematics*, 9, pp.43-73.

Boulding, Kenneth. 1956, *The Image*. Ann Arbor: University of Michigan Press

Brush, Robert O. and Shafer, Elwood L. 1975, Application of a Landscape-Preference Model to Land Management. In Zube *et al.* (eds.) (1975), pp.168-82

Buckley, Frank. 1971, An Approach to a Phenomenology of At-homeness. In Giorgi *et al.* (eds.) (1971), pp.198-211

Buttimer, Anne. 1972, Social Space and the Planning of Residential Areas. *Environment and Behavior*, 4, pp.279-318.

——, 1974, *Values in Geography.* Commission on College Geography Resource Paper No.24. Washington, DC: Association of American Geographers

——, 1976, Grasping the Dynamism of Lifeworld. *Annals of the Association of American Geographers*, 66, pp.277-92.

——, 1978, Home, Reach, and the Sense of Place. In *Regional identitet ouch förändring i den regionala samverkans samhälle*, pp.13-39. Uppsala: Acta Universitatis Upsaliensis Symposia.

Callan, Hilary. 1970, *Ethology and Society.* Oxford: Clarendon Press

Canter, David, and Lee Terrence (eds.). 1974, *Psychology and the Built Environment.* New York: John Wiley and Sons

Carpenter, C.R. 1958, Territoriality: a Review of Concepts and Problems. In *Behavior and Evolution*, eds. A. Roe and G.G. Simpson. New Haven: Yale University Press

Chapin, F.S., and Hightower, H.C. 1966, *Household Activity Systems: a Pilot Investigation.* Chapel Hill, North Carolina: Institute for Research in Social Science, University of North Carolina

Cobb, Edith. 1977, *The Ecology of Imagination in Early Childhood.* New York: Columbia University Press

Coles, Robert. 1967, *Migrants, Sharecroppers, and Mountaineers.* Boston: Little, Brown Co.

——, 1973, *The Old Ones.* Albuquerque: University of New Mexico Press

Cooper, Clare. 1974, The House as a Symbol of Self. In Lang *et al.* (eds.) (1974), pp.130-46.

Cosgrove, Dennis. 1978, Place, Landscape, and the Dialectics of Cultural

Geography. *Canadian Geographer*, 22, pp.66-72.
Craik, Kenneth. 1970, Environmental Psychology. In *New Directions in Psychology*, 4, pp.1-121. New York: Holt, Rinehart and Winston
——, 1975, Individual Variations in Landscape Description. In Zube *et al.*(eds.) (1975), pp.130-50
Dardel, E. 1952, L'Homme et La Terre: Nature de Réalité Géographique. Paris: Presses Universitaires de France
de Grazia, Sebastian. 1972, Time and Work. In *The Future of Time*, eds. H. Yaker *et al.* New York: Doubleday
De Jonge, Derk. 1962, Images of Urban Areas: Their Structure and Psychological Foundations. *Journal of the American Institute of Planners*, 28, pp.266-76
Downs, Roger M. 1970, Geographic Space Perception: Past Approaches and Future Prospects. In *Progress in Geography*, vol.2, pp.65-108. London: Edward Arnold
Downs, Roger M., and Stea, David (eds.). 1973, *Image and Environment: Cognitive Mapping and Spatial Behavior.* Chicago: Aldine
Downs, Roger M. and Stea, David. 1977, *Maps in Minds: Reflections on Cognitive Mapping.* New York: Harper and Row
Durrell, Lawrence. 1969, *The Spirit of Place.* New York: Dutton
Edie, J.M. (ed.). 1964, *The Primacy of Perception and Other Essays.* Evanston, Illinois: Northwestern University Press
Eliade, Mircea. 1957, *The Sacred and the Profane.* New York: Harcourt, Brace and World
Entrikin, J. Nicholas. 1976, Contemporary Humanism in Geography. *Annals of the Association of American Geographers*, 66, pp.615-32
——, 1977, Geography's Spatial Perspective and the Philosophy of Ernst Cassirer. *Canadian Geographer*, 21, pp.209-22
Fischer, Constance. 1971, Toward the Structure of Privacy: Implications for Psychological Assessment. In A. Giorgi *et al*. (eds.) (1971), pp.149-63
Fried, Marc. 1972, Grieving for a Lost Home. In *People and Buildings*, ed. Robert Gutman, pp.229-48. New York: Basic Books
Garfinkel, Harold. 1967, *Studies in Ethnomethodology*. Englewood Cliffs, New Jersey: Prentice-Hall
Giorgi, Amedeo. 1970, *Psychology as a Human Science: A Phenomenologically Based Approach*. New York: Harper and Row
——, 1971a, Phenomenology and Experimental Psychology: I. In Giorgi *et al*. (eds.) (1971), pp.6-16
——, 1971b, Phenomenology and Experimental Psychology: II. In

Giorgi *et al.* (eds.) (1971), pp.17-29
Giorgi, Amedeo, Fischer, W. and Von Eckartsberg, R. (eds.). 1971, *Duquesne Studies in Phenomenological Psychology, vol.1.* Pittsburgh: Duquesne University Press
Giorgi, Amedeo, Fischer, C. and Murray, E. 1975, *Duquesne Studies in Phenomenological Psychology,* vol.2. Pittsburg: Duquesne University Press
Goethe, Johann Wolfgang von. 1952, *Goethe's Botanical Writings,* trans. Bertha Mueller. Honolulu: University of Hawaii Press
——, 1970, *Theory of Colours,* trans. Charles Lock Eastlake. Cambridge: MIT Press
Gottman, Jean. 1973, *The Significance of Territory.* Charlottesville, Virginia: University Press of Virginia
Graber, Linda H. 1976, *Wilderness as Sacred Space.* Washington, DC: Association of the American Geographers
Grange, Joseph. 1974, Lived Experience, Human Interiority and the Liberal Arts. *Liberal Education,* 60, pp.359-67.
——, 1977, On the Way Towards Foundational Ecology. *Soundings,* 60, pp.135-49
Gregory, Derek. 1978, *Ideology, Science, and Human Geography.* London: Hutchison
Guelke, Leonard. 1971, Problems of Scientific Explanation in Geography. *Canadian Geographer,* 15, pp.38-53
Gulick, John. 1963, Images of an Arab City. *Journal of the American Institute of Planners,* 29, pp.179-98
Gutkind, E.A. 1956, Our World from the Air: Conflict and Adaptation. In *Man's Role in Changing the Surface of the Earth,* ed. William Thomas Jr., pp.1-44. Chicago: University of Chicago Press
Hägerstrand, Torsten. 1970, What About People in Regional Science? *Papers of the Regional Science Association,* 24, pp.7-21
——, 1974, The Domain of Human Geography. In *New Directions in Geography,* ed. R. Chorley, pp.67-87. New York: Cambridge University Press
Hall, Edward T. 1966, *The Hidden Dimension.* Garden City, New York: Doubleday
Hardy, Thomas. 1965, *Jude the Obscure.* New York: Houghton-Mifflin
Hart, Roger A. and Moore, Gary T. 1973, The Development of Spatial Cognition: a Review. In Downs and Stea (eds.) (1973), pp.246-88
Harvey, David. 1969, *Explanation in Geography.* New York: St Martin's Press
Hediger, H. 1949, Säugetier-Territorien und ihre Markierung. *Bijdragen tot de Dierkunde,* 28, pp.172-84

Heidegger, Martin. 1962, *Being and Time*, trans. John Macquarrie and Edward Robinson. New York: Harper and Row

——, 1971. Building Dwelling Thinking. In *Poetry, Language, and Thought*, trans. Albert Hofstadter, pp.145-61. New York: Harper and Row

Hilgard, E.R. and Bower, G.H. 1966, *Theories of Learning.* New York: Appleton-Century-Crofts

Hilgard, E.R., Atkinson, R. and Atkinson, L. 1974, *Introduction to Psychology.* New York: Harcourt Brace Jovanovich, Inc.

Hirst, J.R. 1967. Perception. In *Encyclopedia of Philosophy*, vol.6, pp.79-87. New York: Macmillan Co. and the Free Press.

Hockett, C.F. 1973, *Man's Place in Nature.* New York: McGraw-Hill

Hull, Clark. 1952, *A Behavior System*. New Haven: Yale University Press

Ihde, Don. 1973, *Sense and Significance.* Pittsburgh: Duquesne University Press

Ittelson, William H., Proshansky, Harold M., Rivlin, Leanne G. and Winkel, Gary H. 1974, *An Introduction to Environmental Psychology.* New York: Holt, Rinehart and Winston

Jacobs, Jane. 1961, *The Death and Life of Great American Cities.* New York: Vintage

Jager, Bernd. 1975, Theorizing, Journeying, Dwelling. In Giorgi *et al.* (eds.) (1975), pp.235-60

James, William. 1902, *Principles of Psychology*, vol.1. New York: Henry Holt and Co.

——, 1958, On a Certain Blindness in Human Beings. In *Talks to Teachers*, pp.149-69. New York: W.W. Norton and Co.

Josephson, Eric, and Josephson, Mary (eds.). 1962, *Man Alone: Alienation in Modern Society.* New York: Dell

Kaplan, Steve, and Kaplan, Rachel (eds.) (1978). *Humanscape: Environments for People.* North Scituate, Massachusetts: Duxbury Press

Keen, Ernest. 1972, *Psychology and the New Consciousness.* Montery, California: Brooks/Cole Publishing Company

——, 1975, *A Primer in Phenomenological Psychology.* New York: Holt, Rinehart and Winston

Klett, Frank, and Alpaugh, David. 1976, Environmental Learning and Large-scale Environments. In Moore and Golledge (eds.) (1976), pp.121-30

Klopfer, P.H. 1969, *Habitats and Territories: a Study of the Use of Space by Animals.* New York: Basic Books

Koch, Sigmund. 1964, Psychology and Emerging Conceptions of Knowledge as Unitary. In *Behaviorism and Phenomenology*, ed. T.W. Wann, pp.1-45. Chicago: University of Chicago Press

Krawetz, Natalia. 1975, The Value of Natural Settings in Self-Environment Mergence. Paper presented at a symposium, 'Children, Nature, and the Urban Environment', George Washington University, 21 May, sponsored by US Forestry Service

Kwant, Remy C. 1963, *The Phenomenological Philosophy of Merleau-Ponty.* Pittsburgh: Duquesne University Press

Lang, J., Burnette, C., Moleski, W. and Vachon, D. (eds.). 1974, *Designing for Human Behavior: Architecture and the Behavioral Sciences.* Stroudsburg, Pennsylvania: Dowden, Hutchison and Ross

Langan, Thomas. 1959, *The Meaning of Heidegger: a Critical Study of Existential Phenomenology.* New York: Columbia University Press

Lee, Dorothy. 1959, *Freedom and Culture.* New York: Prentice-Hall

Lee, T. 1968, The Urban Neighborhood as a Socio-spatial Schema. *Human Relations*, 21, pp.241-68

Leff Herbert. 1977. *Experience, Environment, and Human Potentials.* New York: Oxford University Press.

Ley, David. 1977, Social Geography and the Taken-for-granted World. *Transactions, Institute of British Geographers*, 2, pp.498-512

Ley, David and Cybriwsky, Roman. 1974, Urban Graffiti as Territorial Markers. *Annals of the Association of American Geographers*, 64, pp.491-505

Ley, David, and Samuels, Marwyn (eds.). 1978, *Humanistic Geography: Problems and Prospects.* Chicago: Maaroufa Press

Leyhausen, Paul. 1970, The Communal Organization of Solitary Mammals. In Proshansky *et al.* (eds.) (1970), pp.183-95

Lindsey, Robert. 1976, Los Angeles Car Habit Hard to Break. *New York Times*, 20 October, p.26

Lloyd, W.J. 1966, Landscape Imagery in the Urban Novel: Sources of Geographic Evidence. In Moore and Golledge (eds.) (1976) pp.279-85

Lorenz, Konrad. 1966, *On Aggression.* New York: Harcourt, Brace and World

Lowenthal, David. 1961, Geography, Experience, and Imagination: Toward a Geographical Epistemology. *Annals of the Association of American Geographers*, 51, pp.241-60.

Luijpen, William A. 1960, *Existential Phenomenology.* Pittsburgh: Duquesne University Press

Lyman, S.M. and Scott, M.B. 1967, Territoriality: a Neglected Sociological Dimension. *Social Problems*, 15, pp.236-49.

Lynch, Kevin. 1960, *The Image of the City*. Cambridge: MIT Press
MacLeod, R.B. 1969, Phenomenology and Crosscultural Research. In *Interdisciplinary Relationships in the Social Sciences*, eds. M. and C.W. Sherif. Chicago: Aldine
Malmberg, T. 1979, *Human Territoriality*. The Hague: Martinus Nijhoff
Maslow, Abraham. 1968, *Towards a Psychology of Being*. New York: Harper and Row
——, 1969, *Psychology of Science*. Chicago: Henry Regheny Co.
McConnaughey, Bayard H. 1974, *Introduction to Marine Biology*. St Louis: C.V. Mosby Co.
Merleau-Ponty, Maurice. 1962, *Phenomenology of Perception*, trans. Colin Smith. New York: Humanities Press
——, 1963, *The Structure of Behavior*, trans. A.L. Fisk. Boston: Beacon Press
Metton, Alain. 1969, Le Quartier: Etude Geographique et Psycho-Sociologique. *Canadian Geographer*, 13, 299-316
Michelson, William. 1966, An Empirical Analysis of Urban Environmental Preferences. *Journal of the American Institute of Planners*, 32, pp.355-60.
Moncrieff, Donald W. 1975, Aesthetics and the African Bushman. In Giorgi *et al*. (eds.) (1975), pp.224-32
Moore, Gary T. 1973, Developmental Variations Between and Within Individuals in the Cognitive Representation of Large-scale Spatial Environments. Unpublished master's thesis, Clark University
——, 1974, The Development of Environmental Knowing: an Overview of an Interactional-Constructivist Theory and Some Data on Within-Individual Development Variations. In Canter and Lee (eds.) (1974), pp.184-94
——, 1976, Theory and Research on the Development of Environmental Knowing. In Moore and Golledge (eds.) (1976), pp.138-64
——. and Golledge, R.G. (eds.), 1976, *Environmental Knowing: Theories, Research, and Methods*. Stroudsburg, Pennsylvania: Dowden, Hutchison and Ross.
——, 1976, Environmental Knowing: Concepts and Theories. In Moore and Golledge (eds.) (1976), pp.3-24
Morris, D. 1968, *The Naked Ape*. New York: McGraw-Hill
Murch, G.M. 1973, *Visual and Auditory Perception*. Indianapolis: Bobbs-Merrill
Nasr, Seyyed Hossein. 1968, *Man and Nature*. London: George Allen

and Unwin, Ltd

Natanson Maurice. 1962. Phenomenology: a Viewing. In *Literature, Philosophy, and the Social Sciences*, pp.3-25. The Hague: Martinus Nijhoff

Newman, Oscar. 1973, *Defensible Space.* New York: Macmillan

Nicholaides, Kimon. 1949, *The Natural Way to Draw: a Working Plan for Art Study.* Boston: Houghton-Mifflin

Norberg-Schultz, Christian. 1971, *Existence, Space, and Architecture.* New York: Praeger

Osgood, C.E. 1953, *Method and Theory in Experimental Psychology.* New York: Oxford University Press

Pastalan, L.A. and Carson, D. (eds.). 1970, *The Spatial Behavior of Older People.* Ann Arbor: University of Michigan Press

Piaget, J. and Inhelder, B. 1956, *The Child's Conception of Space.* New York: Humanities Press

Polanyi, Michael. 1964, *Personal Knowledge.* New York: Harper and Row

——, 1966, *The Tacit Dimension*. New York: Doubleday

Porteous, J.D. 1976. Home: the Territorial Core. *Geographical Review*, 66, pp.383-90

——, 1977, *Environment and Behavior: Planning and Everyday Urban Life.* Reading, Massachusetts: Addison-Wesley

Proshansky, Harold M., Ittelson, William H. and Rivlin, Leanne, G. (eds.). 1970, *Environmental Psychology: Man in His Physical Setting.* New York: Holt, Rinehart and Winston

Rabil, A. 1967, *Merleau-Ponty: Existentialist of the Social World.* New York: Columbia University Press

Rapoport, Amos. 1977, *Human Aspects of Urban Form.* Oxford: Pergamon Press

Relph, Edward C. 1970, An Inquiry into the Relations Between Phenomenology and Geography. *Canadian Geographer*, 14, pp.193-201

——, 1976a, The Phenomenological Foundations of Geography. Discussion Paper No.21, Department of Geography, University of Toronto

——, 1976b, *Place and Placelessness.* London: Pion

Rogers, Carl R. 1969, Towards a Science of the Person. In *Readings in Humanistic Psychology*, eds. Anthony J. Sutuch and Miles A. Bich, pp.21-50. New York: Free Press

Roszak, Theodore. 1969, *The Making of a Counterculture.* New York: Doubleday

——, 1973, *Where the Wasteland Ends.* New York: Doubleday

Rowles, Graham D. 1978, *Prisoners of Space? The Geographical Experience of Elderly People.* Boulder, Colorado: Westview Press

Saarinen, Thomas F. 1969, *Perception of Environment.* Commission on College Geography Resource Paper No.5. Washington, DC: Association of American Geographers

——, 1974, Environmental Perception. In *Perspectives on Environment*, eds. Ian R. Manners and Marvin W. Mikesell, pp.252-89. Washington, DC: Association of American Geographers

——, 1976, *Environmental Planning: Perception and Behavior.* New York: Houghton-Mifflin

Samuels, Marwyn S. 1971, Science and Geography: an Existential Appraisal. Unpublished PhD dissertation, University of Washington

Scheflen, Albert F. 1976, *Human Territories: How We Behave in Space-Time.* New York: Prentice-Hall

Schiffman, Harvey Richard. 1976, *Sensation and Perception: an Integrated Approach.* New York: John Wiley and Sons

Schwenk, Theodore. 1961, *Sensitive Chaos.* London: Rudolf Steiner Press

Seamon, David. 1976a, Extending the Man-Environment Relationship: Wordsworth and Goethe's Experience of the Natural World. *Monadnock*, 50, 18-41

——, 1976b, Phenomenological Investigation of Imaginative Literature. In Moore and Golledge (eds.) (1976), pp.286-90

——, 1977, Movement, Rest, and Encounter: a Phenomenology of Everyday Environmental Experience. PhD dissertation, Clark University

——, 1978a, Goethe's Approach to the Natural World: Implications for Environmental Theory and Education. In Ley and Samuels (eds.) (1978), pp.238-50

——, 1978b, Goethe's Delicate Empiricism: Its Use in the Qualitative Description of Human Experience. Paper presented at the American Psychological Association meetings, Toronto, Ontario, 28 August

——, 1979, Phenomenology, Geography, and Geographic Education. *Journal of Geography in Higher Education* (forthcoming)

Searles, Harold F. 1960, *The Nonhuman Environment.* New York: International Universities Press

Shepard, Paul. 1969, Introduction: Ecology and Man – a Viewpoint. In *The Subversive Science: Essays Toward an Ecology of Man*, eds. Paul Shephard and Daniel McKinley. Boston: Houghton-Mifflin

Skinner, G. William. 1964, Marketing and Social Structure in Rural China. *Journal of Asian Studies*, 24

Slater, Philip. 1970, *The Pursuit of Loneliness: American Culture at the Breaking Point.* Boston: Beacon Press

Smith, Frank. 1975, *Comprehension and Learning.* New York: Holt, Rinehart and Winston

Soja, Edward. 1971, *The Political Organization of Space.* Commission of College Geography Resource Paper No.8. Washington, DC: Association of American Geographers

Sommer, Robert. 1969, *Personal Space: the Behavioral Basis of Design.* Englewood Cliffs, New Jersey: Prentice-Hall

Spickler, Stuart F. (ed.). 1970, *The Philosophy of the Body.* Chicago: Quadrangle

Spiegelberg, Herbert. 1971, *The Phenomenological Movement: an Historical Introduction*, vols. 1 and 2. The Hague: Martinus Nijhoff

Stea, David. 1976. Program Notes on a Spatial Fugue. In Moore and Golledge (eds.) (1976), pp.106-20

Stea, David, and Blaut, James M. 1973, Notes Toward a Developmental Theory of Spatial Learning. In Downs and Stea (eds.) (1973), pp.51-62

Stevick, Emily L. 1971, An Empirical Investigation of the Experience of Anger. In Giorgi *et al.* (eds.) (1971), pp.132-48

Straus, Erwin W. 1966, The Upright Posture. In *Phenomenological Psychology.* New York: Basic Books

Suttles, Gerald D. 1968, *The Social Order of the Slum.* Chicago: University of Chicago Press

——, 1972, *The Social Construction of Communities*. Chicago: University of Chicago Press

Taylor, Charles. 1967, Psychological Behaviorism. In *The Encyclopedia of Philosophy*, vol.1, pp.516-20. New York: The Macmillan Co. and the Free Press

Thoreau, Henry David. 1966, *Walden and Civil Disobedience.* New York: Norton

Thrift, Nigel. 1977, An Introduction to Time Geography. In *Concepts and Techniques in Modern Geography.* University of East Anglia Geographical Abstracts, No.13

Tolman, C. 1973, Cognitive Maps in Rats and Men. In Downs and Stea (eds.) (1973), pp.27-50; originally in *Psychological Review*, 55 (1948), 189-208

Tuan, Yi-Fu. 1961, Topophilia – or Sudden Encounter with Landscape. *Landscape*, 11, pp.29-32

——, 1965, 'Environment' and 'World'. *Professional Geographer*, 17, pp.6-8

——, 1971a, Geography, Phenomenology, and the Study of Human Nature. *Canadian Geographer*, 25, pp.181-92

——, 1971b, *Man and Nature.* Commission on College Geography Resource Paper No.10. Washington, DC: Association of American Geographers

——, 1974a, Space and Place: Humanistic Perspective. In *Progress in Geography*, vol.6, pp.211-52. London: Edward Arnold

——, 1974b, *Topophilia: A Study of Environmental Perceptions, Attitudes, and Values.* Englewood Cliffs, New Jersey: Prentice-Hall

——, 1975a, Images and Mental Maps. *Annals of the Association of American Geographers*, 65, pp.205-13

——, 1975b, Place: an Experiential Perspective. *Geographical Review*, 65, pp.151-65

——, 1977, *Space and Place: the Perspective of Experience.* Minneapolis: University of Minnesota Press

Vine, Ian. 1975, Territoriality and the Spatial Regulation of Interaction. In *Organization of Behavior in Face-to-face Interaction*, eds. Adam Kenton *et al.*, pp.357-87. The Hague: Mouton Publishers

Von Eckartsberg, Rolf. 1971, On Experiential Methodology. In Giorgi *et al.* (eds.) (1971), pp.66-79

Von Uexkull, J. 1957, A Stroll through the Worlds of Animals and Men: a Picture Book of Invisible Worlds. In *Instinctive Behavior*, trans. K.S. Lashley, pp.5-80. New York: International Libraries Press

Vycinas, Vincent. 1961, *Earth and Gods.* The Hague: Martinus Nijhoff

Wallace, Anthony C. 1961, Driving to Work. In *Context and Meaning in Anthropology*, ed. E. Spiro, pp.277-92. New York: Free Press

Wapner, S., Cohen, S.B. and Kaplan, B. (eds.). 1976, *Experiencing the Environment.* New York: Plenum

Webber, M.M. 1970, Order and Diversity: Community without Propinquity. In Proshansky *et al.* (eds.) (1970), pp.533-49

Webster's Seventh Collegiate Dictionary. 1963, Springfield, Massachusetts: G. and C. Merriam Co.

Webster's Third New International Dictionary. 1966, Springfield, Massachusetts: G. and C. Merriam Co.

Weiss, Peter. 1966, *Marat-Sade*, trans. Geoffrey Skelton. New York: Bantam

Wheeler, James O. 1972, Trip Purposes and Urban Activity Linkages. *Annals of the Association of American Geographers*, 62, pp.641-54

White, Lynn, Jr. 1967, The Historical Roots of Our Ecological Crisis. *Science*, 155, pp.1203-7

Wild, John. 1963, *Existence and the World of Freedom.* Englewood Cliffs, New Jersey: Prentice-Hall

Wisner, Ben. 1970, Protogeography: Search for Beginnings. Discussion Paper for Association of American Geographers' annual meetings, August (mimeographed)

Wolf, Thomas. 1973, *You Can't Go Home Again.* New York: Harper and Row

Wordsworth, William. 1936, *Wordsworth: Poetical Works*, ed. Thomas Hutchison. London: Oxford University Press

Wright, J.K. 1947. Terrae Incognitae: the Place of Imagination in Geography. *Annals of the Association of American Geographers*, 37, pp.1-15

Zaner, Richard M. 1971, *The Problem of Embodiment: Some Contributions to a Phenomenology of the Body.* The Hague: Martinus Nijhoff

Zeitlin, Irving M. 1973, *Rethinking Sociology.* New York: Appleton-Century-Crofts

Zube, Ervin H. 1973, Scenery as a Natural Resource. *Landscape Architecture*, 63, pp.126-32

Zube, Ervin H., Brush, Robert O. and Fabos, Julius G. (eds.). 1975, *Landscape Assessment.* Stroudsburg, Pennsylvania: Dowden, Hutchison and Ross

Zube, Ervin H., Pitt, David G. and Anderson, Thomas W. 1975, Perception and Prediction of Scenic Resource Values of the Northeast. In Zube *et al.* (eds.) (1975), pp.151-67

INDEX

activity spaces 33, 88
Adams, J.S. 36
Adams Area, Chicago 88
Allard, 72n2
Allport, F.H. 100, 102n3
Alpaugh, D. 51
Appalachia 136
Appleyard, D. 36n4
appropriation (and at-homeness) 78, 80-1, 87, 88-9, 95
Ardrey, R. 72n2
at-easeness (and at-homeness) 78, 83-4, 87, 89, 95
at-homeness 70, 77, 86-93, 101, 132, 160, 161; and dwelling 92-6; and encounter 107, 117-20; and place ballet 95-6; components of 78-85, 87; definition 70; its development 89-90, 150-1
attachment (feeling-subject) 75-6
attachment (place ballet) 150-1
attention, selective 24
attitudes toward nature 16, 92, 93, 96; and environmental education 113, 124-7, 156-8
attraction (place ballet) 144-5
authenticity 118, 140-1; and places 141-4
awareness continuum 101, 102, 115

Bachelard, G. 19n4, 67, 134
Backster, C. 99
Bannan, J.F. 48, 50, 52n1
Banse, E. 99
Barral, M.R. 52n1, 116
basic contact 115-17; *see also* encounter, perception, movement
Beck, R. 36n4, 51
behavioural geography *see* geography, behavioural
behaviourism 34, 35, 36-7; critique of 43, 51, 121-2; interpretation of body 42, habit 39-40, learning 50-1, perception 100, 121
Bennett, J. G. xi, 137n1, n2, 138
Berry, W. 91-2, 93, 95, 96, 159
Blaut, J. 37n6, 51
Boal, F. 72n1, 88
body ballet 54-5, 60-1, 79, 143; definition 54
body-subject 35, 54, 76, 77, 79, 143; and body ballets 54-5; and habit 40-1; and immersion-in-world 161-2; and Merleau-Ponty 46-8; and time-space routines 55-6; definition 41; learning for 48-50
Bollnow, O. 60
Boorstin, D. 63
Bortoft, H. xi, 22
Boulding, K. 36n4
Brush, R.O. 122, 123
Buckley, F. 19n4
Buttimer, A. 10, 19n4, 20, 21, 35, 72n1, 88, 131, 151-2

Callan, H. 72n2
care 59, 85, 92-6 *passim*, 107, 125, 157
caretaker 92
Carpenter, C.R. 89
Carson, D. 72
centring (in plan ballet) 145
centres 73-4, 132
channelling (in place ballet) 145
Chapin, F.S. 36
Clack University, Worcester, Ma. 25
Cobb, E. 114
cognition, spatial 33, 34, 36, 37, 39, 42; critique of 43, 51-2, 160-2; interpretation of body 42; interpretation of habit 39; role in daily movement 43-5, 62-3, 160
cognitive maps 34, 39, 43, 51-2, 160
cognitive theories: and spatial behaviour 33, 34, 36n3, n4; interpretation of habit 39; approach to body 41-3; and perception 121; chief limitation 160-2; *see also* behaviourism
Cohen, S.B. 10, 18n

Coles, R. 15, 54, 136
coloured shadows 108-9, 113
comfortableness 147-8
common man 140-1, 153
community 9, 63-5, 90-1, 94-6 *passim*, 144, 151-2
Cosgrove, D. 19n4
Craik, K. 18n1, 122, 124, 127n3, n4
Cybriwsky, R. 72n1

Dardel, E. 19n4
defensible space 64-5, 88
de Grazia, S. 63, 99
de Jonge, D. 36n4
delicate empiricism 124-7, 151
distinctiveness (plan ballet) 148-9; authentic 148; inauthentic 148
diversity (place ballet) 145-6; primary 146; secondary 146
Downs, R.M. 34, 36n1, 3, 39, 100, 162n3
Duquesne University, Pittsburgh, Pa. xi
Durrell, L. 98
dwelling 92-6, 126, 127, 160; and at-homeness 92-3; and design 95-6; and rest 132-7 *passim*; definition 92; dwelling and building 93-4; education for 94-5, 124-31; *see also* rest
dyad 137n2

ecology, dividend, 126-7
ecology, foundational 126-7
education, environmental, 124-7, 151-2, 156-8
Eliade, M. 19n4, 99, 114n2
encounter 17, 28, 97-102, 103, 131, 160, 161; and at-homeness 117-20; and environmental education 124-7, 156-8; and perception 97-102, 121-2; definition 99; in triad of environmental experience 131-2; observations on 183-90
Entrikin, N. 19n4
environmental experience groups 18, 24-8, 29; as a learning process 30, 153-8, 191-200; composition 25-6; organisation 201-2; themes 27-8 202-10
environmental psychology *see* psychology, environmental

epoche 20-1; *see also* phenomenology
everyday environmental experience 15-16
explanation 156; impact on education 156-7; *see also* understanding

feeling-subject 75-7, 161, 162; and immersion-in-world 161-2; definition of 76
Fischer, C. 17, 19n4, 21
Fourfold 92
French Quarter, New Orleans 148
Fried, M. 61-2

generalisation 50-1
geographical world 15
geography: as a discipline 15-16, 69, 71; behavioural 16, 17, 18, 159-62

Giorgi, A. 18n3, 20, 36n5
Goethe, J. 13, 113n1, 128n6; and interpersonal corroboration 21-2; delicate empiricism 125-6, 127, 157
Golledge, R.G. 18n1, 36n1,n3, 45n4
Gottman, J. 71, 72n1
Graber, L. 19n4
Grange, J. 29, 92, 113, 126-7, 156-7
Greenwich Village, New York 58
Gregory, D. 19n4, 162n1
group inquiry 21-4, 153-6
Guelke, L. 162n1
Gulick, J. 36n4
Gutkind, E.A. 99

habit 38-43, 60-2, 160; behaviourist interpretation 39-40; cognitive interpretation 39; learning of 48-50; Merleau-Ponty's interpretation 46-8; *see also* body-subject, movement
habituality 118-20; triad of 139, 141-2, 153
Hägerstrand, T. 36n1
Hall, E. 72n1
Hardy, T. 31
Hart, R. xi, 36n3, 51, 52
Harvey, D. 162n1
Hediger, H. 86, 87
Heidegger, M. 10, 74, 85, 86, 92-3, 126, 127, 140-1; and dwelling 92-6

heightened contact 103, 111-13, 115, 124-7
higher contemplation 125
Hightower, H.C. 36n1
Hilgard 39, 50, 51
Hirst, J.R. 100, 102n2
Hockett, C.F. 58
home 78, 131, 132; *see also* at-homeness, dwelling, rest
horizons of reach 131-7 *passim*
Hudson, W.H. 73
Hull, C. 37n6

Ihde, D. 18n3
inauthenticity 118, 140-1; and placelessness 141-2; *see also* authenticity, dwelling
inconspicuous familiarity 74
Inhelder, B. 36n3
insideness 89-90; behavioural 90; empathetic 90; existential 59, 90, 141; *see also* Relph, E.
insider's view 151-2
intersubjective verification 21
invitation (place ballet) 149-50
immersion-in-world 161
Ittelson, W. 18n2, 100, 128n7

Jacobs, J. 59n1, 143, 146, 152n1, n2; and community 64, 65; and diversity 146; and place ballet 58, 64, 95, 144, 145, 151; and watching 106-7
Jager, B. 19n4, 133, 136-7
James, W. 54, 73
Jaspers, K. 162n1
Josephson, E. and M. 63
journey 133, 135, 136-7; *see also* movement, rest

Kaplan, B. 18n1
Kaplan, R and S. 18n1
Keen, E. 18n3, 100, 121
kitsch 64
Klett, F. 51
Koch, S. 36n5
Krawetz, N. 101, 112
Kwant, R. 52n1

landscape assessment 122; critique of 122-4
Langan, T. 120
Lee, D. 113-14n2
Lee, T. 88
Leff, H. 19n1, 36n1
Ley, D. 19n4, 72n1
Leyhausen, R. 86, 87
lifeworld 20, 21, 22, 28, 29; awareness of 153-5
limpets 69
lived-space 60
Lorenz, K. 63, 72n1, 86
Los Angeles, California 61
Lowenthal, D. 19n4, 23
Lynch, K. 36n4

McConnaughey, B.H. 69
MacLeod, R.B. 18n3
Malmberg, T. 72n1
Maslow, A. 112-13
Menomini Indians 57-8
mergence, tendencey toward 101; *see also* encounter, heightened contact
Merleau-Ponty, M. xi, 35, 36, 40, 45, 52, 102; and bodily learning 50; and body-subject 46-8; and perception 116-17, 121; *see also* body-subject, habit
Metton, A. 72n1
Michelson, W. 122
Moncrief, D.W. 19n4, 114n2
Moore, G.T. xi, 18n1, 36n1, n3, 45n4, 162n2
Morris, D. 72n2
movement 17, 28, 33-5, 99, 131; and body 40-1; and body-ballets 54-5; and habit 38-9; and time-space routines 55-6; conventional interpretations 33-5, 42-3; definition 33; dialectic with perception 115-17; dialectic with rest 131-5; interpreted by Merleau-Ponty 46-8; observations on 163-72; *see also* behaviourism, body-subject, cognitive theories, habit, rest
Murch, G.M. 100

Naar, S.H. 114n2
Natanson, M. 20
natural attitude 20, 22, 154; *see also* lifeworld, phenomenology
need 63, 108
Newman, O. 64-5, 72n1, 88
New York City 60
Nicholaides, K. 128n5
Norberg-Schultz, C. 132
noticing 103, 108-11, 115, 124; person-grounded 109-10; world-

grounded 109; *see also* encounter

objective knowledge 21
obliviousness 103, 104-5, 115, 155; *see also* encounter
openness 118-20, 153, and education 153, 151; triad of 139, 153, 157
Osgood, C.E. 36n3
outsideness 89-90; existential 90; objective 90; *see also* place, Relph, E.
outsider's view 151-2

Pastalan, L.A. 72n1
peak experiences 112-13
perception 99-101, 121; and basic contact 115-17; and movement 115-17; interpretations: behaviourist 100, cognitive 100, gestalt 102n3, phenomenological 100-2, 121-2; *see also* basic contact, encounter, movement
phenomenological reduction *see* reduction, phenomenological
phenomenology 16; as an investigatory method 20-4; as a perspective iv, 16-17, 18, 131-2; definition 16; value to geography 16-18, 159-62
Piaget, J. 36n3
place ix, 64, 65; and community ix, 64, 95-6; and dwelling 92-6 *passim*; and insideness 59, 89-90; and locality 9, 90-1, 95-6, 142; and place ballet 56-9, 64-5, 143-4 *passim*; and placelessness 64, 141-2, 158; as a function of territoriality 70-1; authentic 141-2, 143; definition 64; fragmentation of 9, 63, 91, 141-2; inauthentic 141-2; past and present 9, 90-1; planning for 151-2; *see also* placelessness, Ralph, E.
place ballet 54, 57-9, 143-4; and community 63-5; and education 151-2, 153, 158; and existential insideness 56; definition 56; dimensions of: attachment 150-1, attraction 144-5, comfortableness 148-9, diversity 145-6, fostering 63-5, 151-2
placelessness 64, 91, 97, 141-2, 158; *see also* place, Relph, E.
places for things 28, 74-5

Polanyi, M. 53n2
Porteous, J.D. 18n1, n2
positivism 159-60, 162n1,n2
privacy 21, 80, 81
psychology, environmental 16, 159, 160-2

Rapoport, A. 18n1,n2
reduction, phenomenological 20 *see also* phenomenology
regeneration 78, 81-3, 87, 89; *see also* at-homeness, rest
regions 74
reinforcement 39; *see also* behaviourism
relatedness of self 161
relatedness with the environment 112
relocation, neighbourhoods 61-2; *see also* community, place
Relph, E. 10, 19n4,n5, 99, 138, 161; authenticity and inauthenticity 118, 125, 141-2; dialectic between inside-outside 134; dwelling 93; imbalance, movement and rest 135-6; insideness 59, 89-90, 145; outsideness 90; place: definition 64, destruction of 65; placelessness 91, definition 64; topophobia 88-9; *see also* phenomenology, place
rest 17, 28, 68-71, 99, 131; and at-homeness 78; and centres 73-4; and feeling-subject 75-7; and places for things 73-4; definition 69-70; dialectic with movement 131-5; imbalance with movement 135-7; observations on 172-83; *see also* at-homeness, movement
Rogers, C. 21
rootedness 78, 79-80, 86-8, 95, 117; *see also* at-homeness, rest
Roszak, T. 13, 63, 99, 112, 121, 124
Rowles, G.D. 19n4

Saarinen, T. 18n1, 62
Samuels, M. 63, 159, 160, 162n1
Scheflin, A.F. 72n1
Schiffman, H.R. 102n4
Schwenk, T. 99, 113, 127
Seamon, D. 19n4, 22, 99, 113, 128n6

Searless, H.F. 99, 112
selective attention *see* attention, selective
separation, tendency toward 101; *see also* encounter
Shafer, E.L. 122, 123
Shepard, P. 161
Skinner, G.W. 152n1
Slater, P. 63, 64, 95
Smith, F. 51
Soja, E. 70-1, 72n1,n2
Sommer, R. 71, 72n1
sparing and preserving 92-3; *see also* Heidegger, M.
spatial behaviour 16, 33-5, 38-45; *see also* behaviourism, body-subject, cognitive theories, habit, movement
Spickler, S.F. 53n2
Spiegelberg, H. 18n3, 20
Stea, D. 34, 36n1,n3, 37n6, 39, 51, 162n3
stimulus-response 34, 37; *see also* behaviourism
Straus, E.W. 53n2
Strindberg 99
strips, roadside 148; *see also* place, placelessness
subjective knowledge 21
suburbs 86, 93, 148; *see also* place, placelessness
Suttles, G. 70n1, 71, 72, 88
synergy (place ballet) 146
systematics 137n1; *see also* Bennett, J.B.

Taylor, C. 36n5
technique 64
territoriality 16, 69, 70-1, 155-6, 160; and at-homeness 86-9; definition 69, 70; in animals 86-7; *see also* rest
tetrad 137n1, 138; of environmental experience 138-9
Thoreau, H.D. 33
Thrift, N. xi, 36n1
time geography 33, 36n1
time-space routine 54, 55-56, 79; definition 55; *see also* body-subject, habit movement
Toffler, A. 63
Tolman, E. 36n3, 37n6
topophilia 88
topophobia 88
triad 131, 132, 137n1, 139; of environmental experience 131-2; of habituality 139, 141-2, 153; of openness 139, 153, 157
Tuan, Y. xi, 19n4, 45n5, 53n2, 88, 112, 129

van Eyck, A. 132, 210
Vienna, Austria 63
Vine, I. 72n1
Von Eckartsberg 22-3, 24
Von Uexküll, J. 86
Vycinas, V. 140, 141

Wallace, A.C. 42
Wapner, S. 18n1
warmth 78, 84-85, 87, 95; *see also* at-homeness, care, dwelling
watching 103, 105-7, 115, 124; and place ballet 106-7; *see also* encounter
water, appreciation of 127
Webber, M. 96
Weiss, P. 153
West End, Boston 61-2
Wheeler, J. 36n1
White, Jr, L. 99
Wild, J. 29
wish 63, 108
Wisner, B. 19n4
Wolfe, T. 78
Worcester, Ma. 24-5
Wordsworth, W. 98
Wood, D, 36n4, 51

Zaner, R.M. 47, 52n1, 53n2, 60, 116
Zeitlin, I.M. 18n3, 21
Zube, E.H. 100, 122-4, 127n2